FORMS FOOTINGS FOUNDATIONS FRAMING

STAIR BUILDING

DONALD R. BRANN

SIXTH PRINTING — 1978
REVISED EDITION

Published by
DIRECTIONS SIMPLIFIED, INC.
Division of
EASI-BILD PATTERN CO., INC.
Briarcliff Manor, N.Y. 10510

Library of Congress Card No. 70-105687

FIRST PRINTING

REVISED EDITIONS
1974, 1975, 1977, 1978

NOTE
Due to the variance in quality and availability of many materials and products, always follow directions a manufacturer and/or retailer offers. Unless products are used exactly as the manufacturer specifies, its warranty can be voided. While the author mentions certain products by trade name, no endorsement or end use guarantee is implied. In every case the author suggests end uses as specified by the manufacturer prior to publication.

Since manufacturers frequently change ingredients or formula and/or introduce new and improved products, or fail to distribute in certain areas, trade names are mentioned to help the reader zero in on products of comparable quality and end use. The Publisher

ESCAPE TO SUCCESS

As an ambitious kid, the youngest of six in a fatherless home, I learned early what it meant to be poor and how work, regardless of what needed to be done or the wages paid, provided an instant cure to hunger and cold. We didn't have the time or choice to consider whether the job was boring or beneath us. All we wanted to do was earn enough to buy coal at $5.00 a ton and eat regularly. In those days only beggars, bums, hoboes and the lazy lived off charity.

My first job paid 25¢ an hour. I was twelve and my favorite story was still "Aladdin and His Magic Lamp." While I always realized that Aladdin was a magical myth, I discovered many years later a way of life that lends considerable authenticity to the tale.

Like most of the poor who struggle for success, I failed many times before I discovered that everyone who lights a lamp over a workbench during every spare hour, can make like magic. A need to make home repairs and improvements economically sparked the idea behind Easi-Bild full-size patterns and home improvement books. Solving a personal problem helped me help millions solve theirs.

Those who transform time into building labor can make many dreams, even good health, come true. This book is a case in point. Read it. See how you can become an expert on forms, footings, foundations and framing. And how doing this work provides relaxation, exercise, escape from tension, while it opens doors to new business opportunities.

Don R. Brann

TABLE OF CONTENTS

DISCOVER GOLD

Relatively few people realize that spare time invested in building labor provides the quickest way to create a sizeable tax-free investment. Every hour you invest in improving or building living space provides big savings. Building labor is not only among the highest paid in the nation, but is also in short supply. When you shell out eight to sixteen dollars an hour for a skilled craftsman, only a small part of what you pay is for labor. You also pay that man's taxes, social security, union dues, transportation, etc. What the craftsman keeps is small compared to what you paid out.

Buying the necessary materials, then investing spare time in following directions you understand, can create an investment equal to thousands of dollars. And the improvement, like real estate, increases in value in direct relation to the rise in inflation, and/or the scarcity of good housing. Each hour invested in building labor generates capital gains when you sell.

There is another financial aspect worth consideration. When you help a skilled craftsman, you gain many benefits. You not only learn how a pro does the work, but you also help set a pace that generates more production per hour. Some people consider stalling an honest way to steal, a form of legal larceny no one says anything about.

Transforming spare time into building labor offers great personal satisfaction, sizeable financial rewards, plus substantial physical and mental benefits. The exercise is extremely beneficial once a person conditions his body to accept the effort. The work is satisfying and is something any healthy person can do, and do well at every age. No one is ever too old to learn. While it takes some people in the sixties a bit longer to do certain jobs, age does not affect the quality of work or lessen the benefits earned. Before starting, read this book through carefully. Study each illustration to familiarize yourself with procedure.

Building anything, whether it's a dog house, lean-to, house, garage, additional space, etc., requires drawing straight lines, sawing along a drawn line, then driving nails. Everyone, even those in the lowest income level, who start from scratch, can practice with low cost tools.

Sawing a short piece of scrap lumber following straight lines, and nailing the pieces together provides practice in sawing and nailing. Those who want to learn construction fast should build a scale model using scale size lumber made from ¼'' plywood or balsa wood. Make or buy scale lumber to this approximate size:

2 x 4 — ¼ x ⅝''; 2 x 6 — ¼ x ⅞''; 2 x 8 — ¼ x 1¼'';
2 x 10 — ¼ x 1½''; 2 x 12 — ¼ x 1¾''.

Actually going through each step of construction as described in this book, learning what goes where and when, enables an amateur to intelligently make application as a helper on any construction job. Being familiar with construction helps move a beginner up the ladder. Helping to clear a site prior to putting in footings and foundation, mixing cement, stacking and carrying block, handling lumber as needed, making saw cuts or driving nails are all stepping stones to new careers.

Housing has always been and will continue to be one of the world's greatest problems. Any way you can get into the act, through new construction or rehabilitation of existing buildings, puts you into the mainstream and provides an opportunity of growing up into a highly profitable and urgently needed consumer service.

Since everyone interested in building must own land, the introductory pages cover some of the practical and legal points that should be considered before purchasing a site. If you already own land, pass the advice on to someone who is beginning to look. Helping people help themselves is one of the most rewarding experiences in life.

The basics outlined in this book meet most building codes. Since a building permit is required in most areas, the building department will specify changes required. Those who read but hesitate to do any of the work are rewarded by learning how a job is done, what materials are required and where each is used. This permits a more intelligent appraisal of a contractor's estimate. It also permits buying materials separately from labor. Why be robbed when a little reading can help save hundreds, even thousands of dollars.

Give a child a shovel and if he is old enough to sit, he will start to dig. Without years of apprentice training, the child will attempt to build a castle, a wall or just a hole in the sand. Digging a hole, something every child does naturally and without instruction, is the hardest first step for an adult. The child does it by instinct, the adult considers digging a trench a big deal. Most adults are so brain-washed they think everything anyone does to earn big money requires special skill.

Building forms, laying footings, foundations and framing, is no more difficult than digging a trench. And this is real easy if you use proper tools, Illus. 1.

TOOLS REQUIRED

The following tools will be needed. Pick, shovel, mason's hoe, cement mixing tub, measuring box for sand, trowel, float, ball of building line, plumb bob, line level, carpenter's level, 50 ft. steel measuring tape, framing square, miter square, hammer, brace and bits, chalk line, and folding zig zag rule. To build forms you will also need a crosscut saw or electric handsaw and a wrecking bar, Illus. 1.

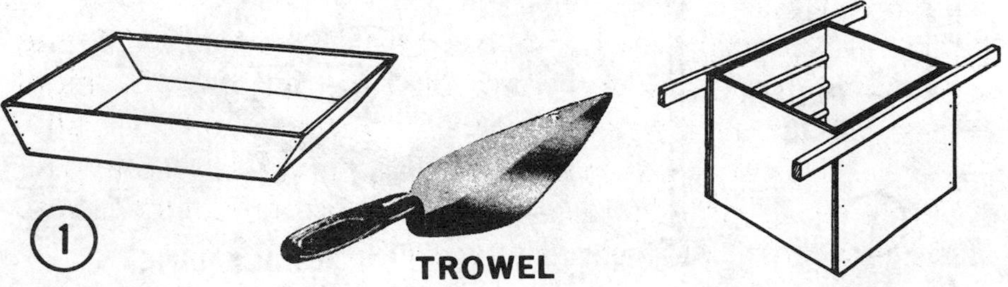

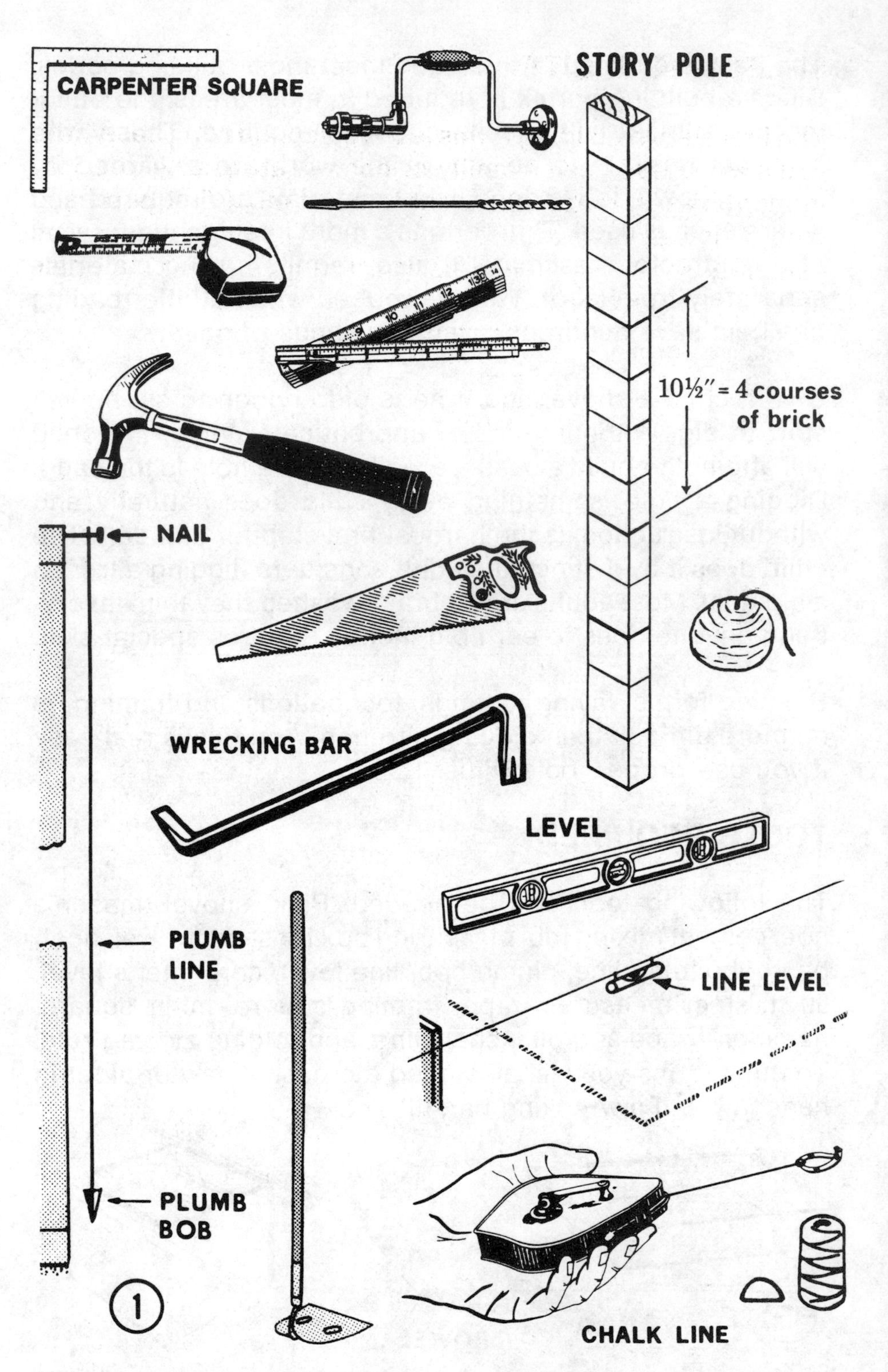
CARPENTER SQUARE
STORY POLE
10½" = 4 courses
of brick
NAIL
WRECKING BAR
LEVEL
PLUMB
LINE
LINE LEVEL
PLUMB
BOB
1
CHALK LINE

SELECTING A SITE

The first step in building forms is the selection of a site. Where you start to build can simplify or complicate your entire job. In many cases, those who want to build must first purchase land. This may be level, hilly, rocky or swampy. In most areas you must obtain a building permit. These require a plot plan Illus. 2, showing location of proposed building.

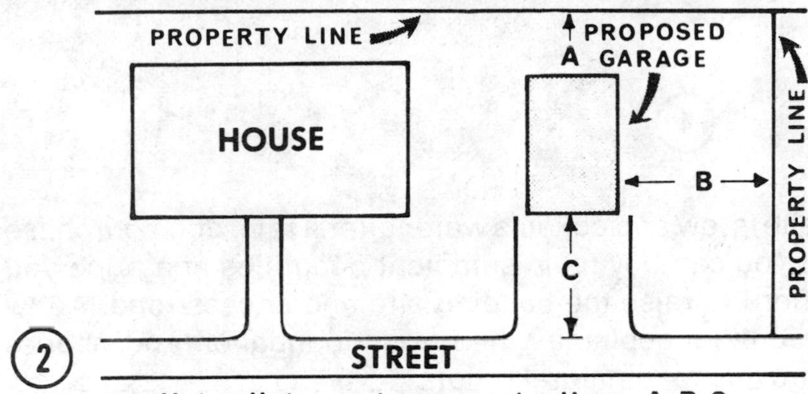

Note distance to property lines-A, B, C

If the site selected is in a low area, a full basement or crawl space, Illus. 3, may not be feasible. In this case, a slab on grade, Illus. 4, may provide the answer. A slab on grade refers to a concrete deck on top of a level bed of gravel or compacted soil, or over a rocky site; or to filling in a low spot with sufficient gravel or compacted soil so that the site is raised sufficiently above grade to eliminate flooding. A slab should be 8″ to 12″ above grade to permit water to run off and not in.

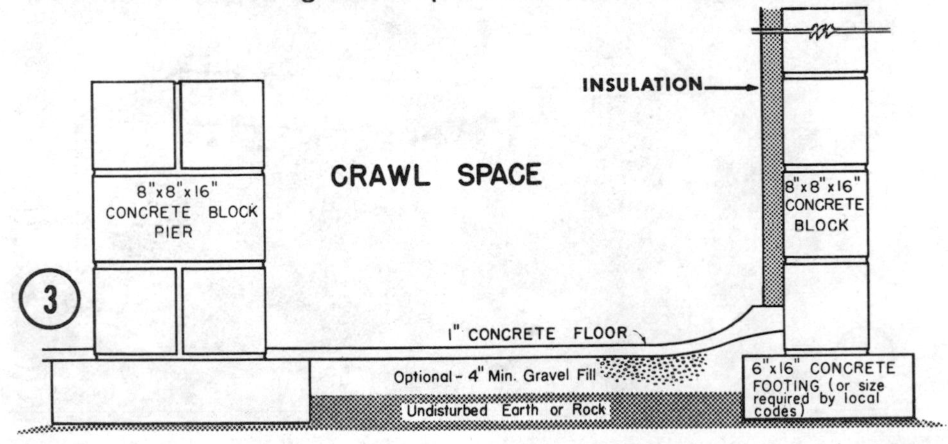

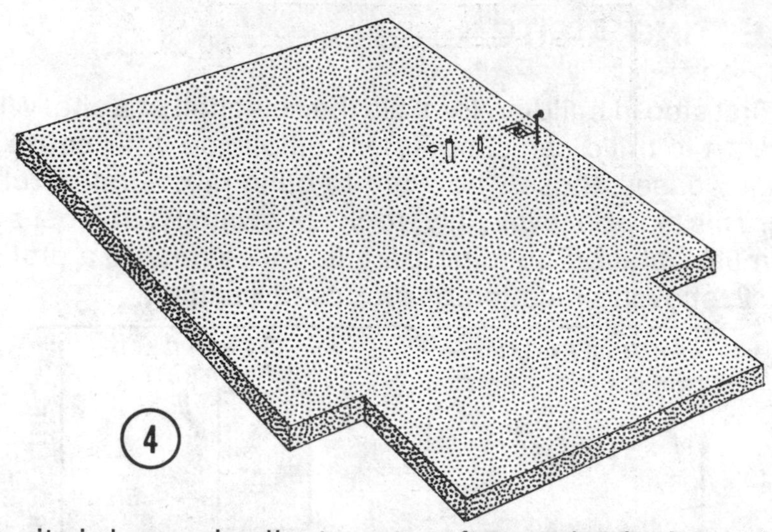

If the site is low and collects water after a rain, don't purchase unless you can buy fill in sufficient quantities at a price you can afford to raise the building site and access road. A low site is only acceptable when your foundation and access road provide for a natural runoff.

If the site is on a steep slope, and only part of it is needed for a house and garage, consider building a cantilever or post-supported structure, Illus. 5. By anchoring framing to rock and to supporting columns, a sizeable house can be built on a hilly or rocky site. If you study the site and rent a jack hammer, Illus. 6, to drill holes, this approach can provide as much house as you want to build at relatively low cost. That is, if the rock pile can be purchased at an acceptable price.

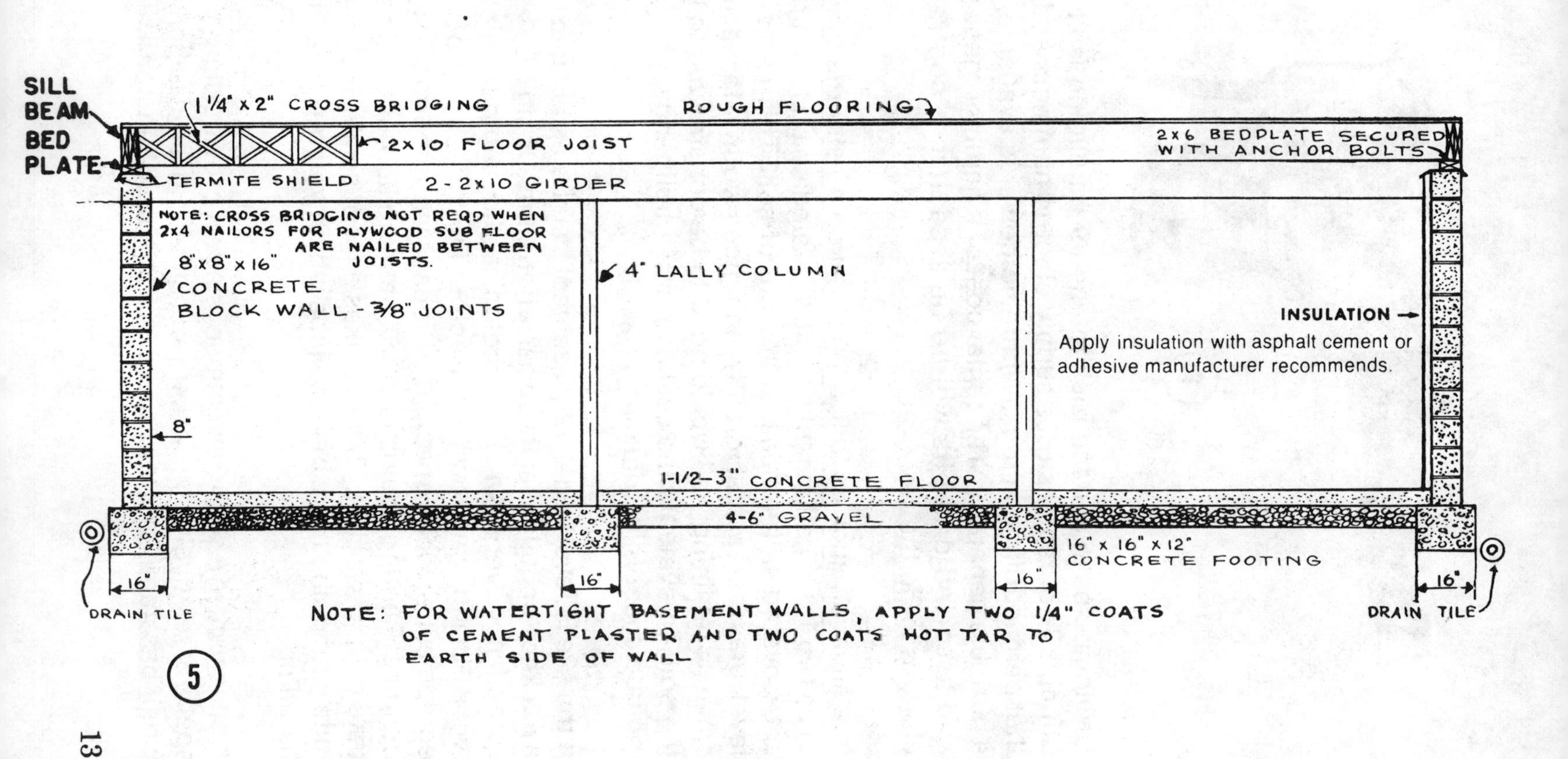
SILL
BEAM
BED
PLATE
1 1/4" x 2" CROSS BRIDGING
ROUGH FLOORING
2x10 FLOOR JOIST
2x6 BEDPLATE SECURED
WITH ANCHOR BOLTS
TERMITE SHIELD
2 - 2x10 GIRDER
NOTE: CROSS BRIDGING NOT REQD WHEN
2x4 NAILORS FOR PLYWOOD SUB FLOOR
ARE NAILED BETWEEN
JOISTS.
8" x 8" x 16"
CONCRETE
BLOCK WALL - 3/8" JOINTS
4" LALLY COLUMN
INSULATION
Apply insulation with asphalt cement or
adhesive manufacturer recommends.
8"
1-1/2–3" CONCRETE FLOOR
4-6" GRAVEL
16" x 16" x 12"
CONCRETE FOOTING
16"
16"
16"
16"
DRAIN TILE
DRAIN TILE
NOTE: FOR WATERTIGHT BASEMENT WALLS, APPLY TWO 1/4" COATS
OF CEMENT PLASTER AND TWO COATS HOT TAR TO
EARTH SIDE OF WALL
5

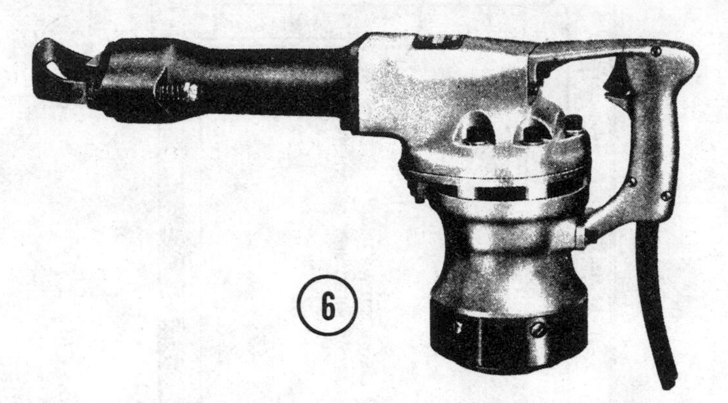

Drilling holes in surface rock to anchor a building isn't very difficult. Tool rental stores supply electric hammers and bits that permit drilling in rock. Where no power is available, rent a portable generator. The rental cost of a hammer, generator and assorted drill bits will be far less than a contractor charges to do the work.

In a situation such as this, a 1,000 gallon oil storage tank, or even two 1,000 gallon tanks can be leveled up on top of rock, and covered with a minimal amount of soil. This can add a level grass area that greatly enhances your site. In most cases, a retaining wall would need to be constructed to hold the soil. To evaluate the cost of a site intelligently, you must include cost of fill, retaining wall, etc.

If the site is level and a full basement can be excavated, hire a backhoe or bulldozer if you can afford one. Or dig it by hand, something every American colonist did without thinking twice. Being your own contractor, finding out who rents equipment or does specialized jobs is an important part of building. Renting equipment for X hours to do work that could take days or weeks by hand, makes a lot of sense. It also permits your doing more of those jobs which you are capable of handling.

A basement is a sound investment since it provides important space at relatively low cost. If you can't build a full basement, a half basement or crawl space is far better than none at all.

The square foot cost is usually less than any other in the house. Space below grade costs less to heat, less to cool, and thanks to air conditioning, provides a desirable, and to some extent intrusion-proof, living area the year round.

If your building program is stymied by money or time, consider using the basement as a living unit. With air conditioning, you can convert a basement into living space while you continue to build above ground.

If you consider buying unimproved land, free of building code restrictions, be sure to consider the cost of digging a well as part of the purchase price. If there is no water adjacent to the property, ask neighboring property owners how deep they dug to obtain an adequate supply. Inquire whether they have water every month of the year. Never ask a native if they have water all summer. Many will answer "yes." Later you may find the wells in that area go dry in September, October or some other month.

Always note whether adjacent property owners have one bathroom, or two, a washing machine, etc. If they conserve water, the chances are it's in short supply. Land without water has little or no value for a homestead.

If you find a site in line with what you can pay, and there are no water mains serving the area, talk to at least two different well drillers who work in the area before putting down a deposit. Make inquiry as to depth of other wells in the area. While this doesn't guarantee an equal flow to your site, it does provide some ballpark figures. Always write down every mention of cost per foot through rock, and rate per foot through earth. Request a written confirmation.

If possible, try to visit homes where the well driller has worked to ascertain whether the work was entirely satisfactory. Also find out how much over the original estimate the job cost. While few natives give advice to strangers, if you hire a local lawyer, he could be an excellent source of good information.

After you find an acceptable site and prior to parting with a deposit, discuss the purchase with a local lawyer. Lawyers, like dentists, doctors and other professionals, frequently charge whatever they think the traffic will bear. For this reason, find out how much he will charge to represent you in the purchase of the property if and when you decide to go ahead, and what his charge will be if you decide against purchase. If the lawyer levels with you, he will be glad to give you an outside estimate. If he says he can't tell what the cost will be, find yourself a new lawyer. Tracing a deed and giving advice concerning the purchase of property is not a difficult job to estimate. Discuss the cost of getting title insurance and what title insurance company he recommends.

Contact the title insurance company and find out what lawyer they recommend in the area prior to retaining the first lawyer you talked to. Ask the lawyer if he currently or previously represented the seller. Query him about local well drillers and other craftsmen. Be sure to ascertain who among those mentioned are presently his clients, and/or whether he had any legal business with or against those mentioned. "Getting dry behind the ears," learning who knows who, is important to a newcomer about to make a sizeable investment in the community.

Since buying land is usually a once or twice in a lifetime operation and one that can cost a bundle if you make a mistake, take each step slowly. Think and talk it over. Be doubly sure you are doing business with people who want to do business with you.

Prior to purchase, add lawyers fees, the highest estimate you get, title search and insurance costs, well costs, etc., to the price asked for the land. Then figure out if it is still a good investment. If you double a well driller's estimate, you will probably be closer to the final cost. Drilling a well is one venture that can't be estimated accurately.

When a well digger says he can't tell how much rock he will strike, or how deep he will have to drill for water, he's telling you the truth, even though past performances in the same area provide some indication as to the depth of a water table. The depth of rock at any one level, coupled with the "honesty" of the driller, can create a memorable experience for those operating within limited financial means.

One important point to keep in mind when talking to a well driller is the question of competition. The rig required to drill a well costs money to move. You can't afford to have your driller travel any great distance. For this reason, don't alienate any affections before or during a contract, or you could be out the cost of a well. Other important points to consider, and each should be spelled out, concern the gallons per minute that the driller believes sufficient to fill his contract, the equipment he will furnish, the size of pipe, storage and pressure tank he agrees to install, and the size of pipe he will run into the house. Many drillers base their contract on drilling a well only. They will not make a connection to a house or furnish a pressure storage tank. Be sure your contract mentions the equipment he agrees to use to drill your well, and whether breakdowns in equipment, and/or loss of time due to equipment failure, are on your bill or at his cost. When you get an estimate, go to a dealer and price a pump, storage tank, etc. Be sure to get a price on the size equipment the driller recommends as being adequate for your needs. A contract with a well driller should be reviewed by your lawyer to make certain it covers everything that needs to be done. Digging a well is an experience recommended only to the legally minded individual who places his faith in God and his confidence in an honest lawyer.

If you believe there is sufficient water in the area to drill a well, the next step is to figure out how to run a water line between the well site and proposed dwelling. If there's an abundance of rock where you need to trench, this could require blasting. If you need to blast, most well drillers recommend doing it prior to drilling a well. Since rock conducts cold, it is imperative to use double or triple thick insulation around the water line to insulate it from the rock.

Another way to connect a water line to a house over rock without blasting is by insulating the pipe, then covering it with the depth of soil that the frost level requires. In this case, compare the cost of fill and a retaining wall, if same is needed, with the cost of blasting. In many areas an insulated plastic water line covered with X inches of fill can withstand fairly cold weather.

Equal in importance to having a supply of water is the question of having space available for a septic tank and field. If the site selected is rocky, it doesn't provide much opportunity for installing a tank or laying a field, Illus. 7. The total length of field depends entirely on the size of the tank. The size of the tank is dependent on the number of people occupying the proposed building.

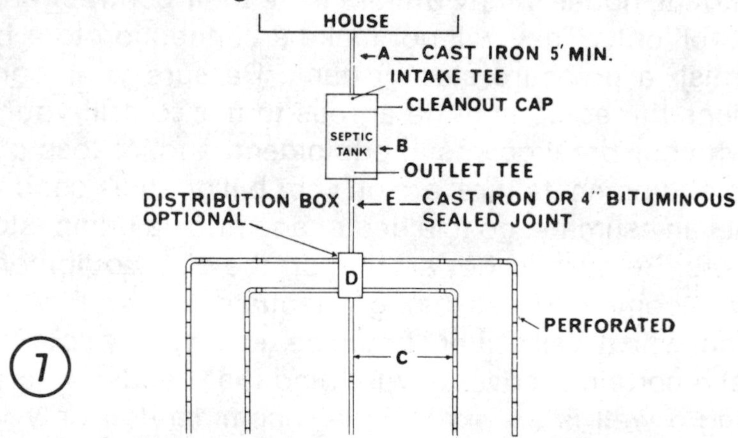

Board of Health departments frequently permit laying an approved length of field to fit space available.

Prior to purchase, it's important to visit the local Board of Health to find out what size field you would be permitted to install to meet local requirements and to be granted an occupancy permit. Since the field and its run off must be located a specified distance from your property line, the property may be worthless as a building site. You might end up using it to park a mobile home, providing there is no local ordinance prohibiting same. If all this sounds discouraging, it is, but it's far better to be disappointed prior to purchase than to wake up some night and realize you've been had.

A normal size septic tank field can require 500 to 600 square feet, 20′ x 30′, even up to 20′ x 50′. And this area cannot be used for any other purpose, even to park a car. If the site contains an abundance of surface rock, you might find it easier to build retaining walls and fill the area with sand or dirt fill to a depth that meets local requirements.

An important condition to this approach is the fact that the runoff from a septic tank field must be absorbed by the ground within your property line and must not drain off into a brook or into neighboring property.

The Board of Health restrictions covering septic tank fields are important to your health and the health of everyone living in the area. Few exceptions are allowed so don't allow a smooth talking real estate agent con you into thinking he can use his influence to obtain a permit when you have been turned down. Even if he could get the permit, living on property where a septic tank starts acting up on a hot summer's day is no fun. Predicate the purchase of the parcel on your being able to obtain both a building as well as a septic tank permit. You need both to build.

If the land appears to be just what you dreamed about buying and there's little or no rock above ground, get permission to drive a crowbar into the ground at various points selected for the house. This will give you some idea how much rock is just below the surface. Do the same in the area selected for a septic tank.

BE SMART

1. Don't pay even a dollar down as a deposit or binder until your lawyer approves the sales contract and you recheck suggestions outlined.

2. Only buy when you can obtain a clear title with no right of way or other encumbrances.

3. Consider whether the area has a future or whether school busing, increased taxes, a proposed highway, or a low-income housing project has recently put a kiss of death on the area.

4. Check to see if the property is protected by zoning that prohibits erection of gas stations, factories, etc.

5. Check whether there are any unpaid taxes, mechanics or other liens against the property. What are present tax assessments? What new assessments have already been adopted that will affect the property? The only place to get a complete answer is at the tax assessor's office. They will also tell you how many times in its recent past the property has changed hands.

6. Is the property convenient to what you need — schools, churches, shopping center, transit facilities, and place of employment? Is there adequate police and fire protection to permit an acceptable insurance rate?

7. Can you build the size house your family requires? Building codes invariably specify how far a house can be constructed from the street or property line. Many people have bought land only to learn they couldn't build the house they wanted because there wasn't sufficient space for a septic tank field.

8. Does the land offer the kind of building site that enhances the style of house you want to build?

9. Does it offer a site with natural drainage away, and not toward the house?

10. Does the site for house, septic tank field and garage still provide sufficient space for outdoor living? Does this space offer the privacy you want? Consider placement of a garage and driveway, its grade and approach to street. Will snow drift into area? Does the site selected for a house allow sufficient area for both a garage and septic tank field?

A local lawyer familiar with the property and its history, may be able to save you much trouble and expense. While most lawyers can't give you an exact cost of a closing, they can tell you the many problems that could arise, and what you can expect in the way of costs. Make arrangements to pay your lawyer one fee if you go through with the closing, another fee if the closing reveals conditions that prevent your taking title. It's always better to back out of a deal with a small loss than to go ahead and take a larger one.

Be sure to have a title search made and buy title insurance. This is a sound investment. Title insurance protects your investment.

Buying a house or land naturally creates fear and packs an emotional wallop. Even a level-headed businessman loses some of his cool when he discovers the amazing number of problems that can arise during a closing. For these reasons, it's important to play safe. Don't pay a dollar down, don't sign or agree to anything until you know what you are buying.

A case in point is the size and shape. Let's say it is offered as an acre. Is the forty odd thousand square feet of land shaped as you think it is, with a wide road frontage, or are you getting fifty foot frontage and an odd-shaped parcel that won't complement your proposed building.

If you don't ask questions, the seller is not obligated to bring them up. Unless you ask, you could one day be surprised to see a bulldozer carve a street through the middle of your property, or a trenching machine digging a ditch for a sewer line where you wanted to locate your living room. Don't take the seller's word on anything unless it's put in writing. If he's honest, he won't hesitate to write down everything he says.

Before buying find out if a lending institution will loan money on the property and how much of a mortgage they'll offer. Their appraisal will give you some inkling as to what others judge its value. Always talk to the most reputable lending institution. Shady loan companies will lend as much money as they can when they investigate your ability to pay. To them the value of the land is incidental.

Psychologists claim physical work helps release tension and a variety of physical activities is important to sound mental health. Many brain shrinkers are dedicated do-it-yourselfers. They advise purchasing carpentry and masonry tools with the same thought and expectation as sporting equipment. Footings and foundations require both concrete and carpentry tools. Illus. 1, shows some of the tools you will be using.

If you decide to mix concrete rather than purchase readymix for footings, you will need a mixing tub, Illus. 8. You can make one using 1 x 12 for sides A, cut to angle shown, plus two ends B. Apply waterproof glue and nail sides to ends. Apply glue and nail a 3/16″ tempered hardboard panel to bottom. Always place tub on a level surface. If you use it where it needs to be supported to make level, nail an extra 3/8″ or 1/2″ plywood panel to bottom as a stiffener. To build a 15-cubic-foot mixing tub 11″ deep, cut A-82″; AA-60″; B-34″.

Paint tub with wood preservative. When thoroughly dry, paint inside surface with used crankcase oil. After using, always scrape out concrete and hose tub thoroughly. When dry, paint with old crankcase oil before reusing.

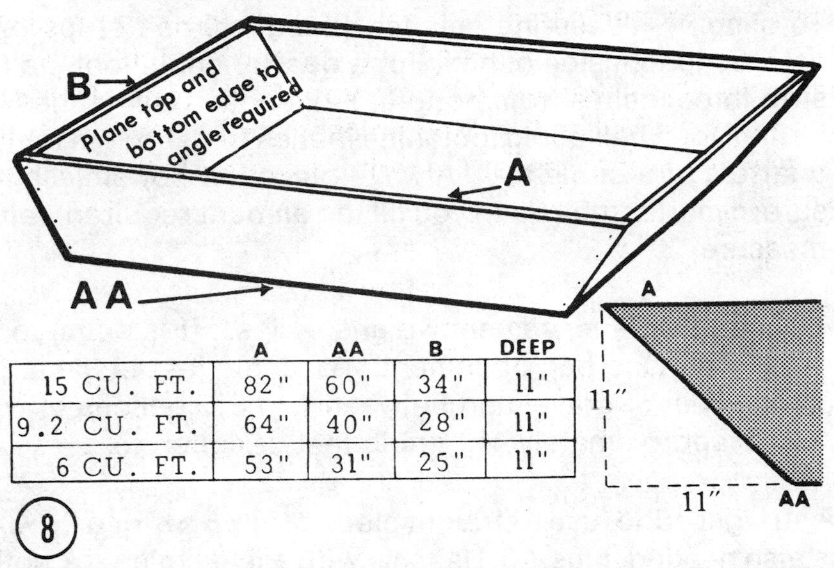

	A	AA	B	DEEP
15 CU. FT.	82"	60"	34"	11"
9.2 CU. FT.	64"	40"	28"	11"
6 CU. FT.	53"	31"	25"	11"

8

To accurately measure ingredients, build a bottomless box, Illus. 9. Cut two pieces of ½″, ⅝″ or ¾″ plyscord, A — 13½″ x 12″; two pieces B — 12″ x 12″. Apply waterproof glue and nail A to B, using 8 penny common nails spaced three inches apart. Cut two 1 x 2 x 24″ for handles. Nail handles to box in position shown. Use a file or rasp to round edges of handles. When filled level with top, measure holds one cubic foot.

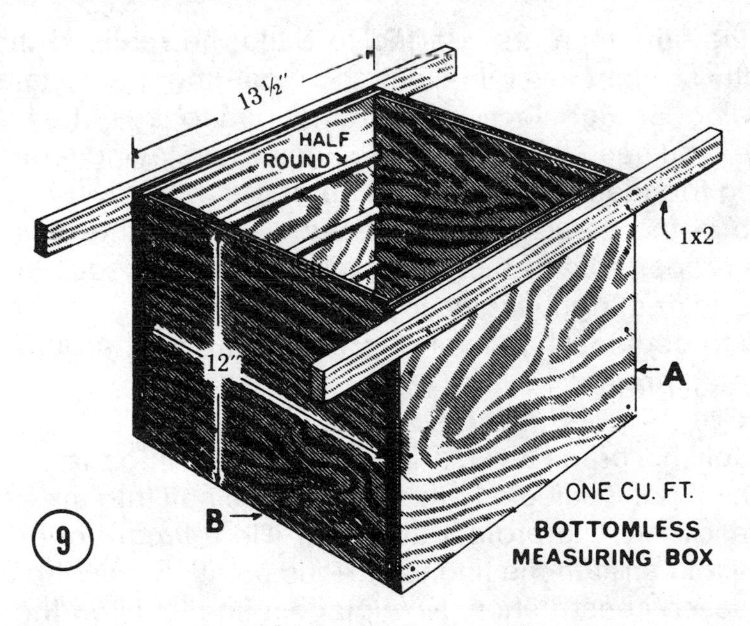

9

BOTTOMLESS MEASURING BOX

To simplify measuring smaller quantities, nail strips of ⅜″ half-round to inside of box. For a quarter cubic foot, nail one strip three inches from bottom; nail another six inches from bottom for a half cubic foot; nine inches from bottom for three quarters of a cubic foot. Always place the bottomless measure in mortar tub. When you fill the amount required, remove measure.

A bag of Portland cement weighs 94 lbs.* It is equal to one cubic foot. One bag of cement, 2¼ cubic feet of sand, plus 3 cubic feet of gravel and between 4 to 5½ gallons of water makes approximately 4½ cubic feet of concrete.

A straight edge (use a straight piece of 5/4 x 6 x 8′ or a 2 x 4 x 8′) is also needed, Illus. 10. Use this with a level to check bottom of trenches, footings, etc.

GUIDE LINES

From the time man first started to build, he realized that a perfectly straight level line was the most important element in building. Straight level lines are needed to lay out a building site. You need the same guide lines when laying footings, building foundation walls, etc. The first step in every building job is to position the proposed structure the distance from a property line which local building codes specify and parallel to or at right angles to that property line. The second step is to decide what level of ground you want around the completed project.

To establish a grade level, drive a stake flush into the ground at the high point selected, Illus. 11. Drive a nail into top of the stake allowing it to project about 1″. Tie a length of line to the nail and stretch the line to various points within the area of proposed construction. Now place a line level on the line

*87.5 lbs. in Canada.

and note the low and high points. This will give you some idea where to remove soil, where to fill in, where to bury a fuel tank, etc.

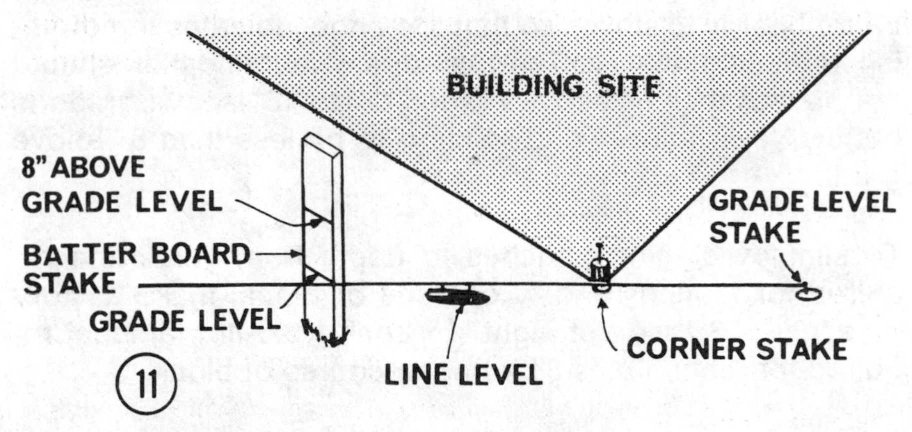

Next drive a stake into ground to indicate a front corner of the proposed building. This corner stake, can project 3″ or 4″ above ground, Illus. 12. Drive a nail into top of this stake.

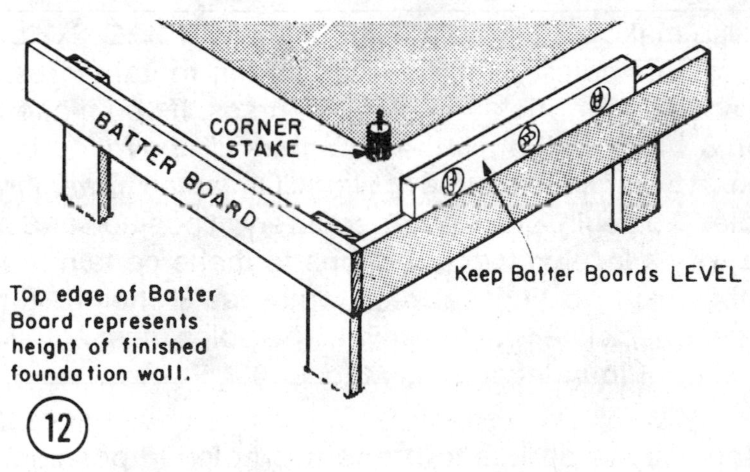

Approximately three feet from the corner stake, drive batter board stakes in position indicated. Use 1x4 sharpened at one end. Batter board stakes can project above batter boards or be sawed off flush with top edge of batter boards. Stretch a line from the grade level stake to the batter board stake. Attach a line level to line. When line indicates level, mark stake, Illus. 11. This mark now indicates grade level on batter board stake.

Measure up 8″, if you want one course of block to show above grade; 16″, if you want two courses above; 24″ for three courses. Drive two more stakes in position shown. Then nail batter boards to stakes so that top edge indicates top of finished foundation wall. The top edge of a foundation should not be less than 8″ above grade, 12″ to 16″ above grade is better. Never allow wood framing to be less than 8″ above grade.

To simplify digging trenches to depth from guide line required for footings and X courses of block, make a story pole, Illus. 13. Use a straight 5/4 x 2 or 5/4 x 3. Allow amount required for footing plus 8″ for each course of block.

After footings have been laid, saw pole at footing mark and use as a guide to check height of footing form from guide line.

You can make a story pole using a straight 5/4x2, 2x2 or 2x4. Professional poles, Illus. 14, available in rental stores, indicate spacing for brick and block courses. If you make poles for an 8″block wall, draw the first mark 8″ above footing. Add marks at 8″ for each course of block. Or fasten a spacing tape to pole. To begin, place poles securely in position and check each with a level in two directions to make certain they are plumb. Fasten a line between poles at 8″ height for first course. Check line with line level. The poles must equal overall height of foundation plus at least 8″.

Another way to do it is to brace two poles in position, then mark the height of first course on pole. When the first line is level, additional 8″ courses are marked off. The story pole not only provides level guide lines but should also indicate window and door openings, lintels, etc.

Illus. 15, shows how a pro sets up poles for outside and inside corners. Make inquiry at your tool rental or masonry supply store. These poles are well worth the cost of rental.

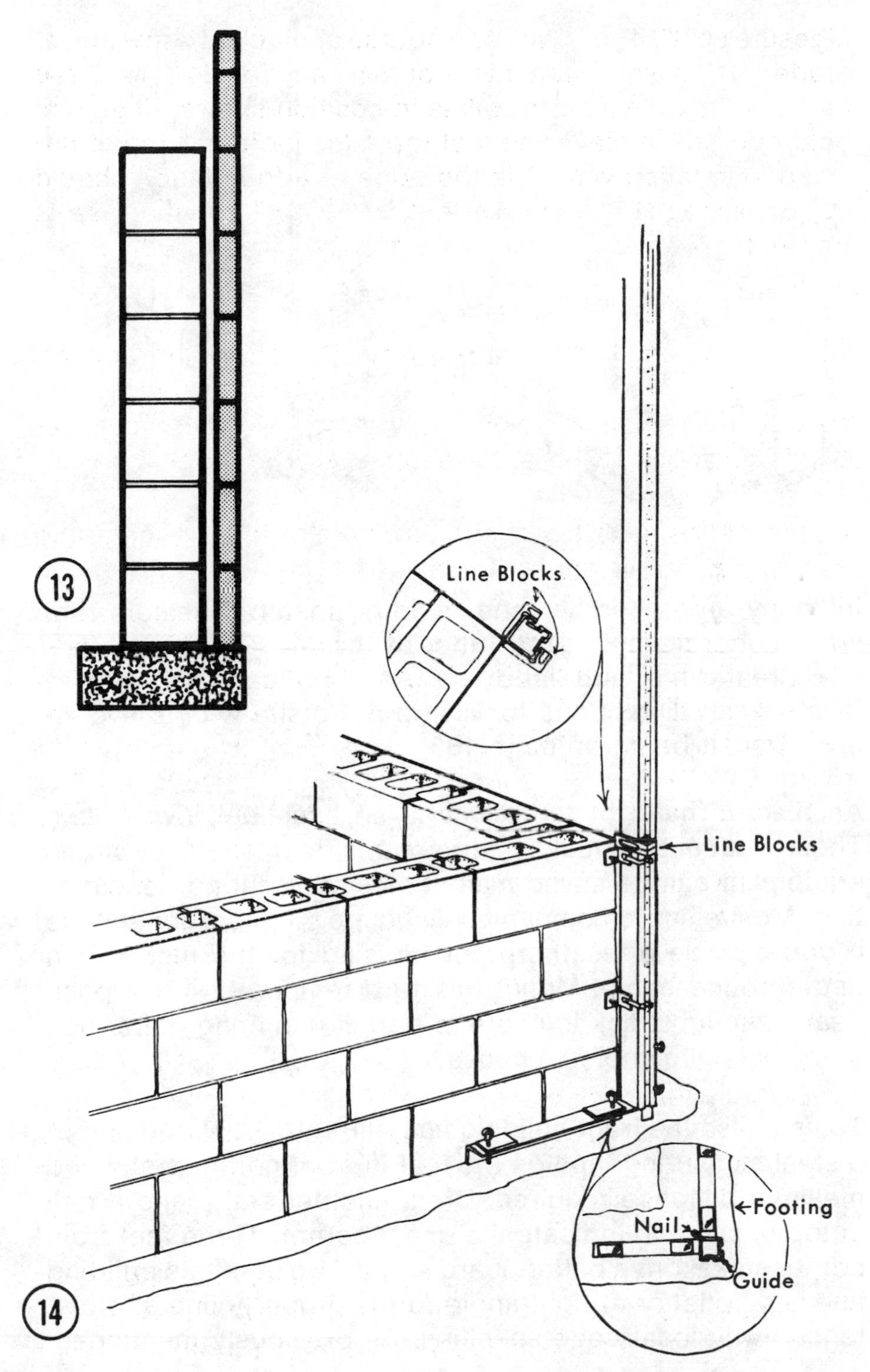
13
Line Blocks
Line Blocks
Footing
Nail
Guide
14

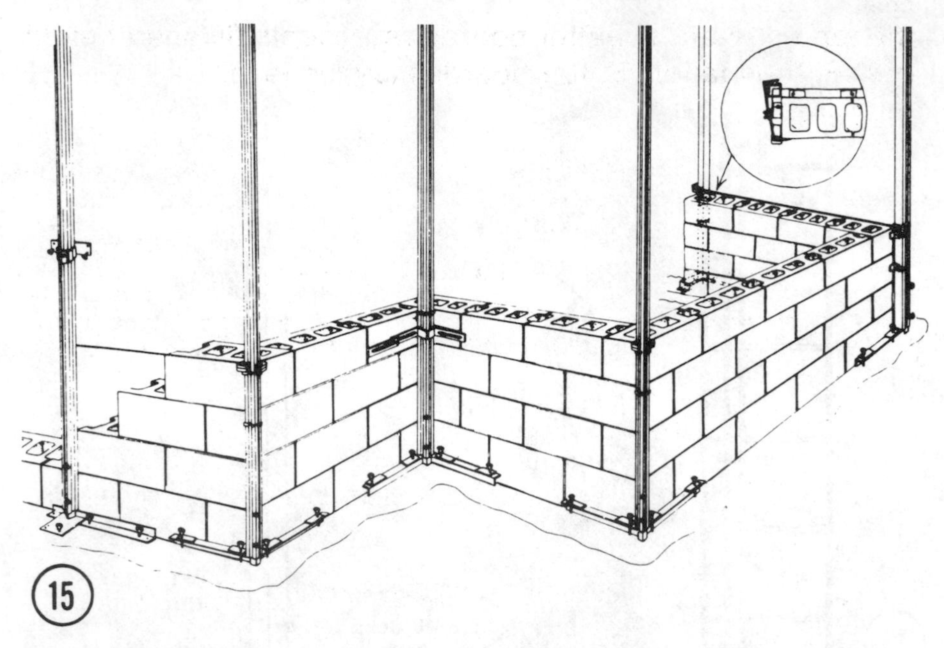

In many areas, lumber and building material dealers rent small concrete mixing machines by the day. These machines are great time and labor savers. You can also rent a level-transit. Directions for laying out a site with a level or level-transit begin on page 163.

An electric hand or table saw is an excellent investment. These frequently equal an extra man's time. The cost of this equipment can be saved many times over during construction. Most electric companies will supply temporary service if you provide a weatherproof housing for the meter. Use underground cable. Mount this to a temporary 4 x 4 post. Leave enough slack in cable to permit mounting meter box on outside of completed house.

To establish the front building line, Illus. 16, hook the end of a steel measuring tape on the nail in front corner stake and measure distance required. Drive another stake and a nail in top of stake to indicate the exact corner. Three feet from corner stake, drive batter board stakes. Be sure this building line is parallel or at right angle to the property line and distance away local code specifies. As previously mentioned, the top edge of the batter boards represents finished height of your foundation. Batter boards must be level.

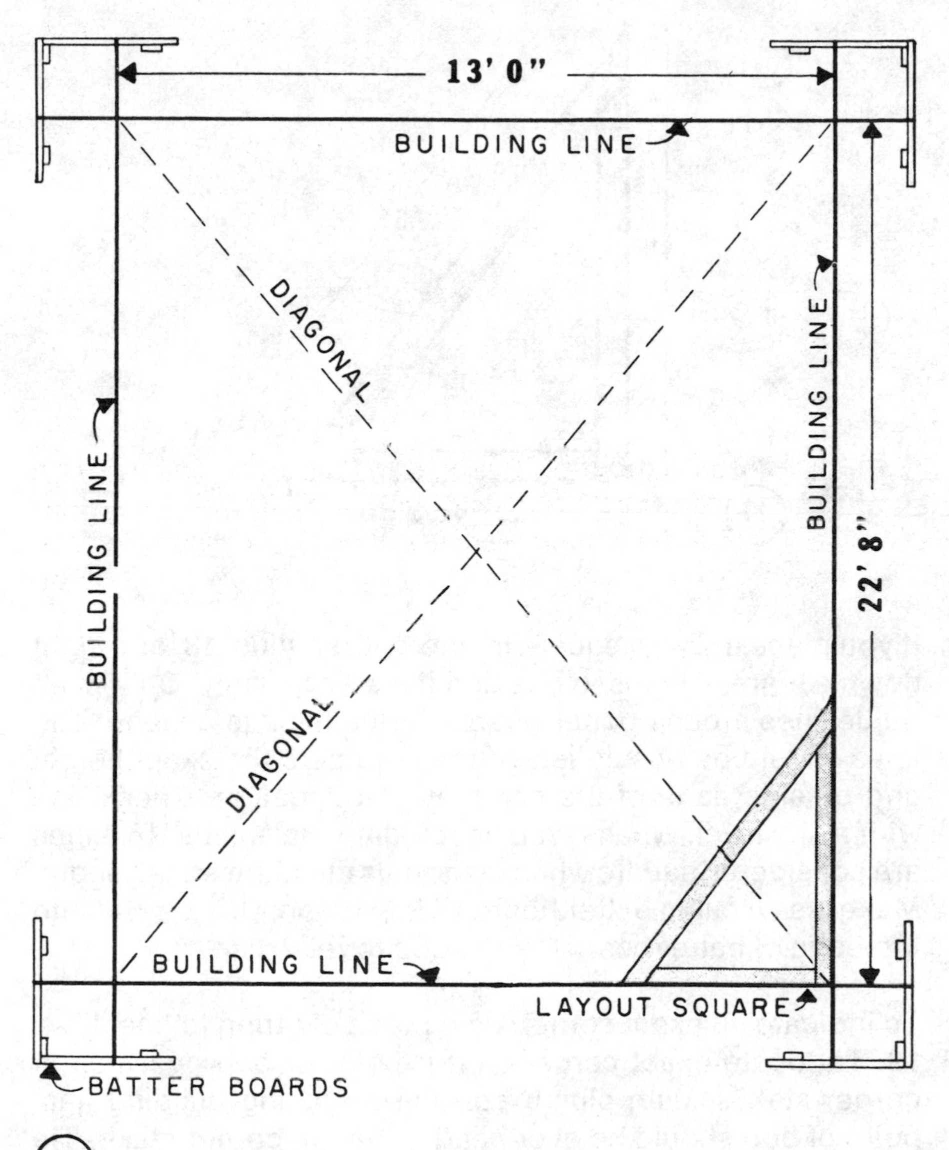

A layout square, Illus. 17, helps to square up guide lines. To make one, mark 3′ on one 1x4, 4′ on another, 5′ on a third. Nail or screw boards together at right angle.

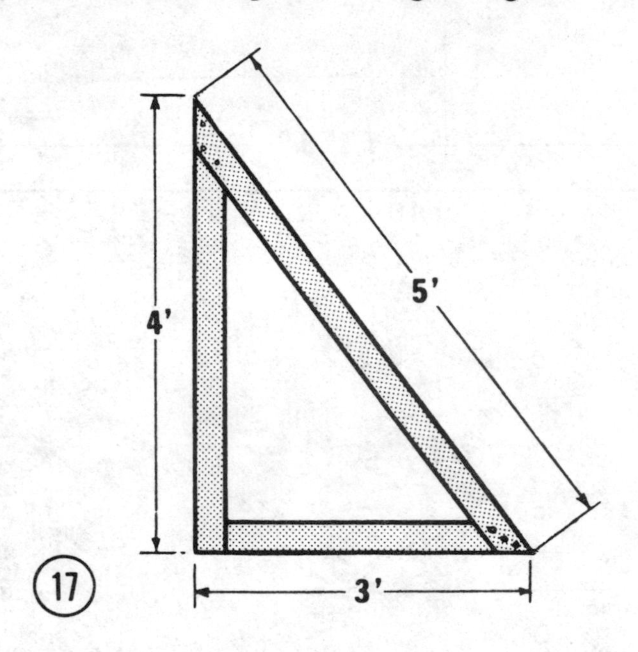

If you place a layout square in one corner, Illus. 18, and sight down square, it helps to position lines accurately. Only wrap guide lines around batter boards, or tie lines to a stone. The lines should be positioned so they indicate the exact height and outside edge of the completed foundation. Check line with a line level. When level, check diagonals, Illus. 16. Lines are considered square when diagonals measure equal length. Make a saw cut in batter board to lower line. Drive a nail into top edge of batter board if you need to raise line.

To indicate an exact corner, tie a piece of string to line, Illus. 19. To locate exact corners, drop point of bob over nail in corner stakes. With plumb bob line touching building line, point of bob should be over head of nail in corner stake. Tie a piece of string to guide line to indicate exact corner. When lines are square and level, pull taut and tie to batter boards at exact height required.

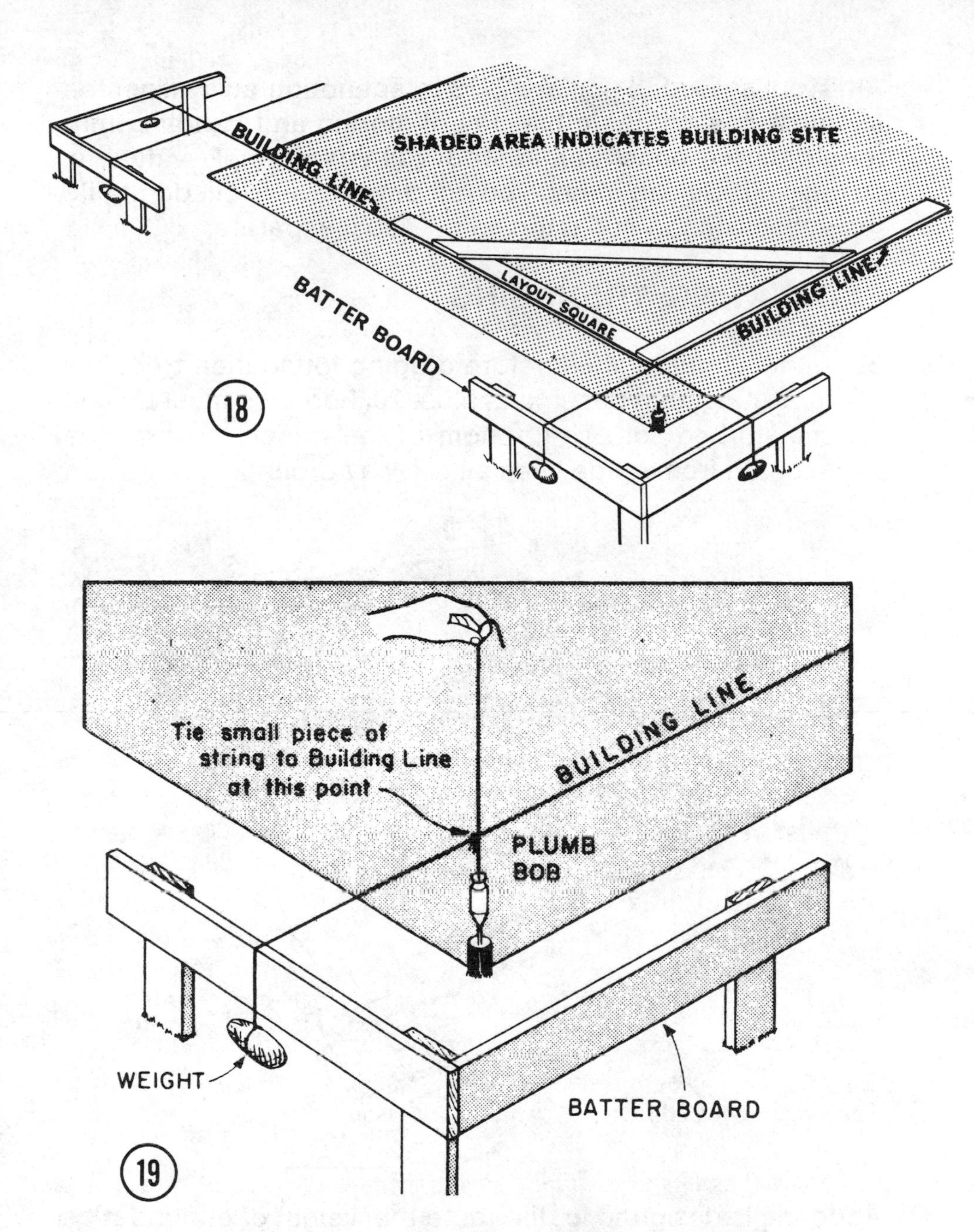

The success of a footing and foundation depends on the guide lines. Unless these are taut, level and square, you can't expect to lay straight footings or build a straight foundation wall. Use guide lines when excavating for the trench. Check depth of trench all the way around against guide lines.

Always remove top soil and pile it far enough away from the site so it doesn't get in the way of storing and handling materials during construction. Top soil is extremely valuable, so save all you can. When you get down to subsoil, don't pile the subsoil on top of the top soil. Keep it separate.

EXCAVATING, TRENCHING

Set up level guide lines before digging foundation trenches or excavating for a basement, Illus. 20. Footings must be laid on undisturbed soil. The bottom of the trench must be the same depth from guide lines all the way around.

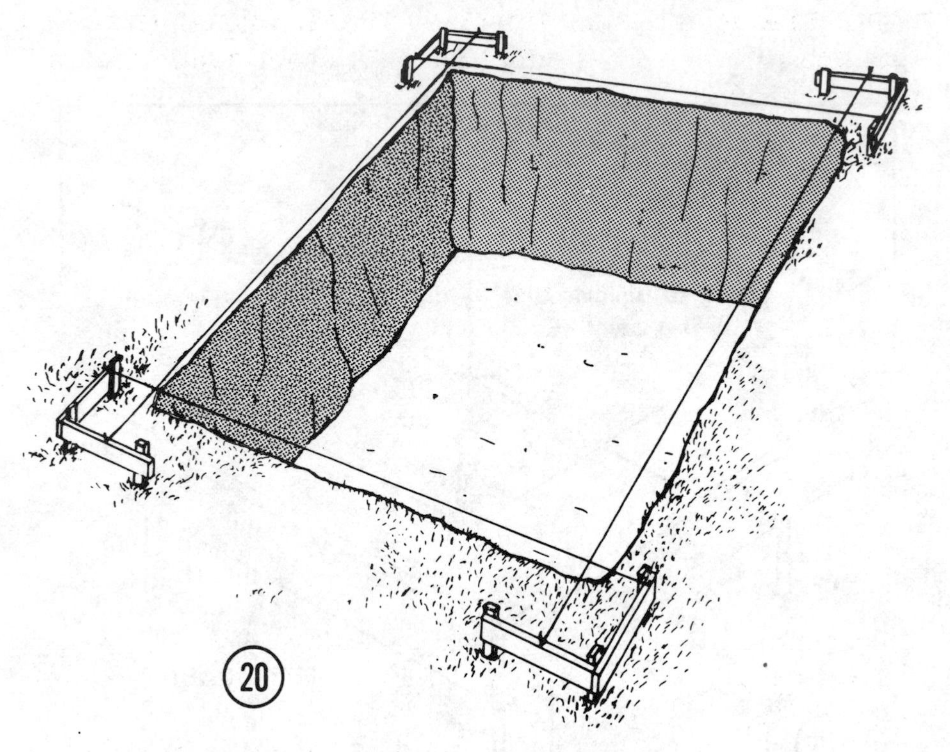

A footing is designed to distribute the weight of a foundation wall and load above. By spreading the load over a wider base than the foundation, greater stability is achieved. The National Building Code recommends footings be placed 1′0″ minimum below frost level. This requires as much as 4′0″ below grade in Boston and Chicago, 1′6″ in Seattle. Consult and follow local building codes. This is important since a freeze-up can crack a footing and cause considerable damage to a foundation.

Builders frequently level a site, then excavate a two-foot-wide trench for 16″ wide footings that require two or more feet of depth. In areas where the thermometer doesn't go below 32°, codes frequently permit trenching to width foundation requires, and using the trench as a form, Illus. 21. Keep sides of trench straight. Use a plumb bob from guide lines to position trench. Check depth of footing to make certain it's equal distance from guide lines all the way around. Fill trench with concrete to equal height from guide lines. This construction requires a level site at equal distance from guide lines all the way around.

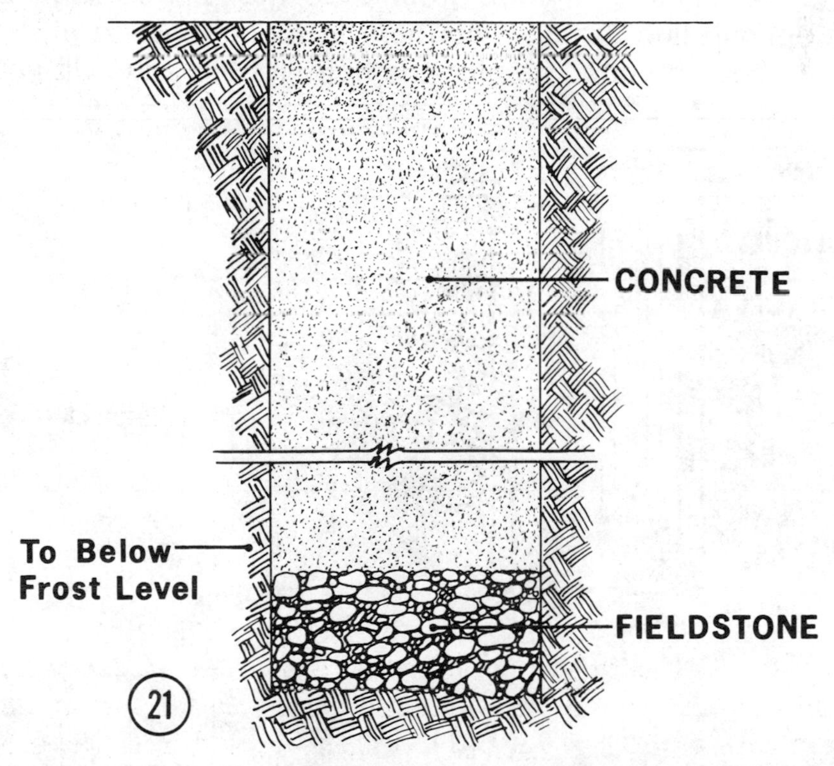

Prior to pouring a footing, it's important to ascertain whether waste and water lines will go under footing or through the foundation wall. When lines are to be installed below footing, make an open end box, Illus. 22, using 1 x 6 or 2 x 8 x 16″ (for a 16″ footing). Place this in position required and build footing form on top. After footings have been allowed to set three days, knock out form. Be sure to locate position of opening on your plan.

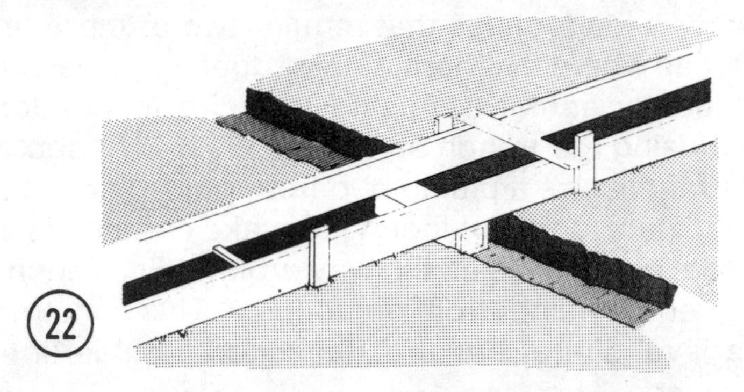

Plan on running an underground electric service cable and telephone line through this hole in footing. When all lines have been roughed in, plug the hole at both ends with concrete.

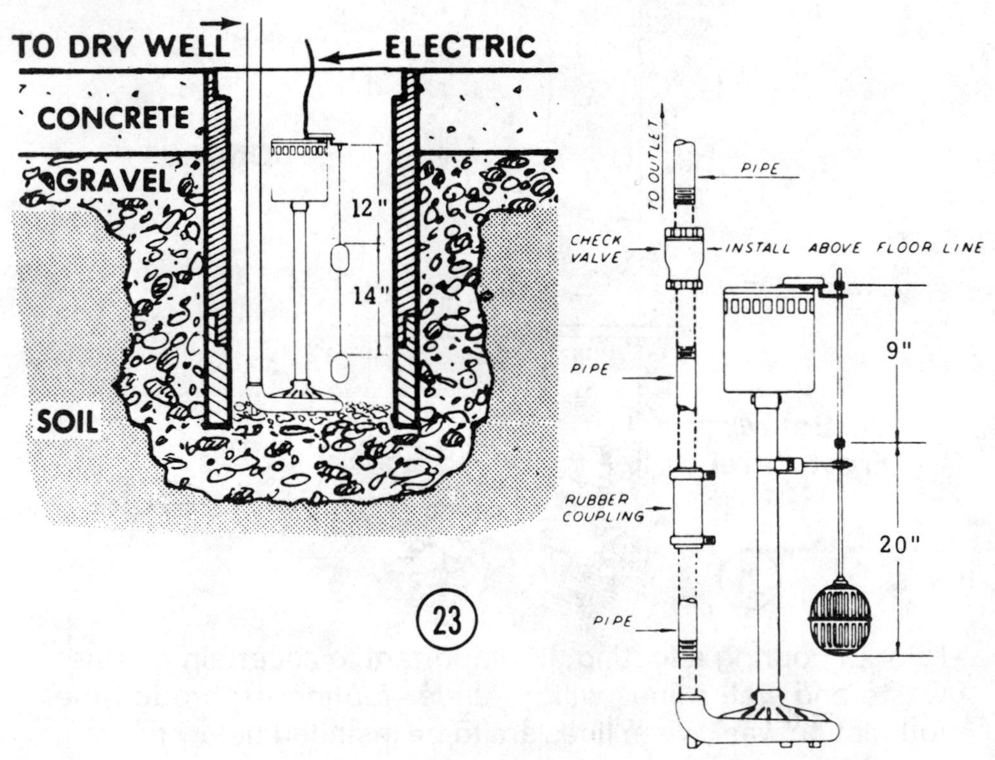

If you plan on installing a sump pump, Illus. 23, in a basement or crawl space, bury a 12″ diameter x 2′0″ tile, Illus. 24, flush with floor, or use tile-size sump pump manufacturer recommends. Tile should be positioned on a bed of gravel. While the waste line from a sump pump can be run through a foundation wall, a better installation can be made by digging a trench and burying the pipe before pouring footings. Plastic pipe is now recommended for a sump pump. Run waste line from sump pump to a dry well, Illus. 25, some distance from and at a depth lower than basement floor.

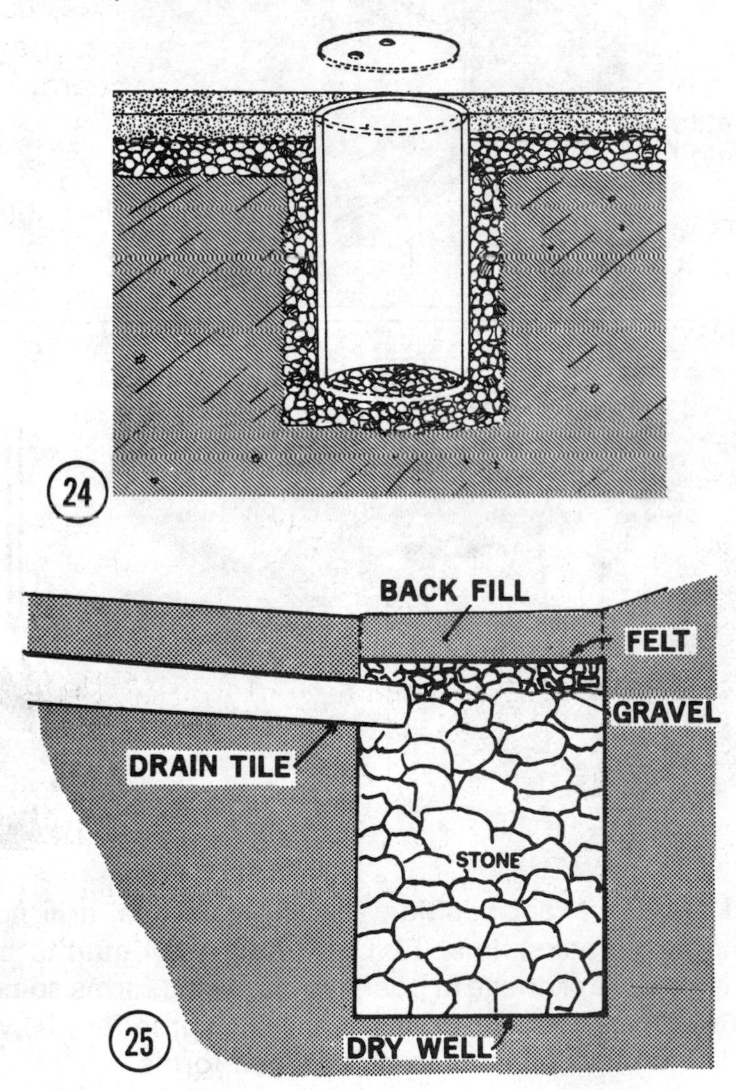

FOOTING FORMS

When footings are laid on grade, codes permit fastening framing directly to footings. In this case use 2 x 8, 2 x 10, or 2 x 12 footing forms to raise footing an acceptable height above grade. Embed anchor bolts, Illus. 26, or anchor clips, Illus. 27, in position that plans specify.

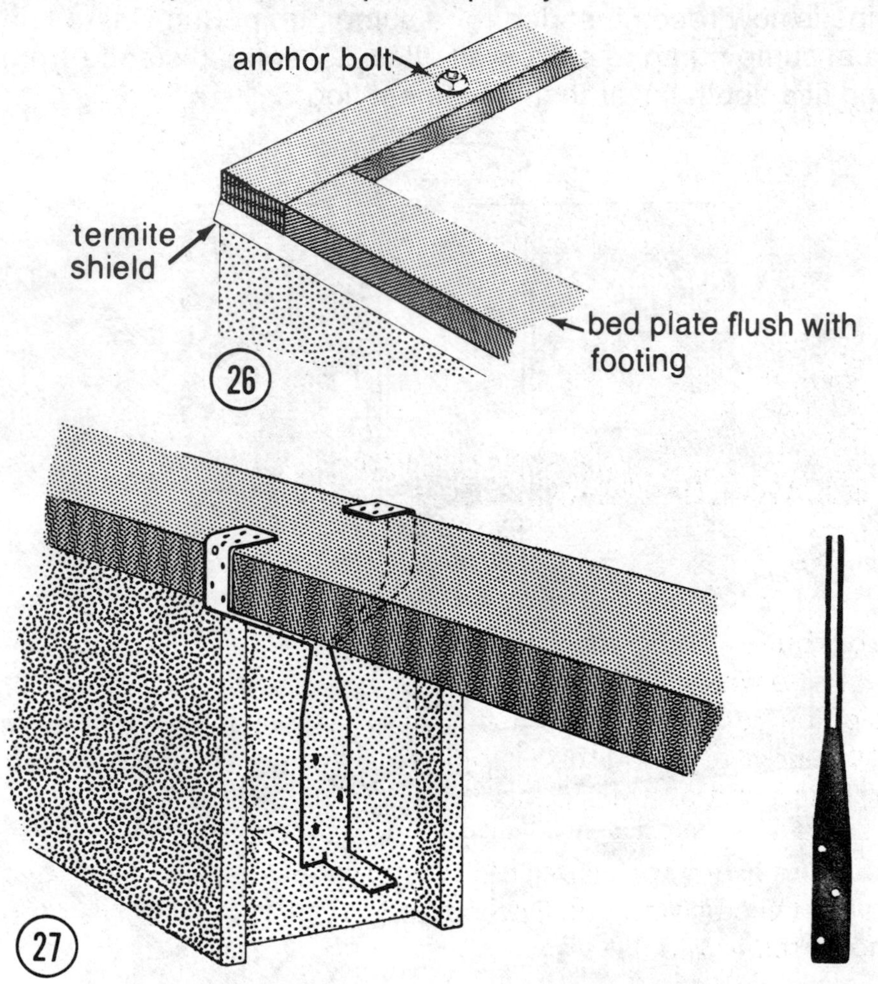

To hold bolts in exact position framing requires, drill holes through 2 x 4 blocks, Illus. 28. Use thickness equal to bed-plate, Illus. 26. Tack these in position to footing forms so bolts protrude between studs. Note how 1 x 2 spreaders A, wire and nails can also be used to build small forms.

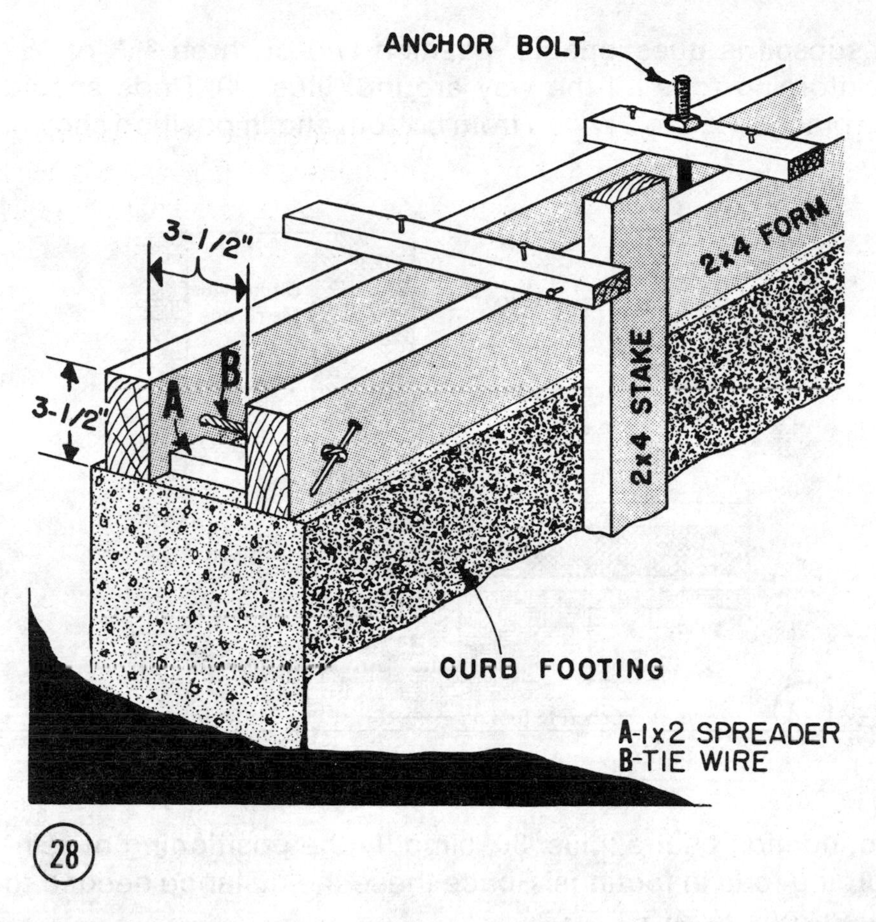

28

Most builders use ½″ x 10″ or ½″ x 12″ bolts with two washers. One washer is embedded against head of bolt, the other is placed on top of bedplate, Illus. 26. Always embed bolts in top course of foundation blocks when building a block wall, or in top of a poured foundation.

Aluminum anchor clips, Illus. 27, can be used in place of bolts. These come flat. Spread and position them before filling cores in blocks.

Drill holes in a piece of 2 x 4 to hold bolts in position, Illus. 28, while filling a block course or form.

Use one part cement to three parts sand to five parts gravel for footings, or buy readymix.

If subsoil is questionable, position two or three 3/8″ or 1/2″ reinforcing rods all the way around, Illus. 29. Rods should be placed 1 1/2″ to 2 1/2″ up from bottom and in position shown.

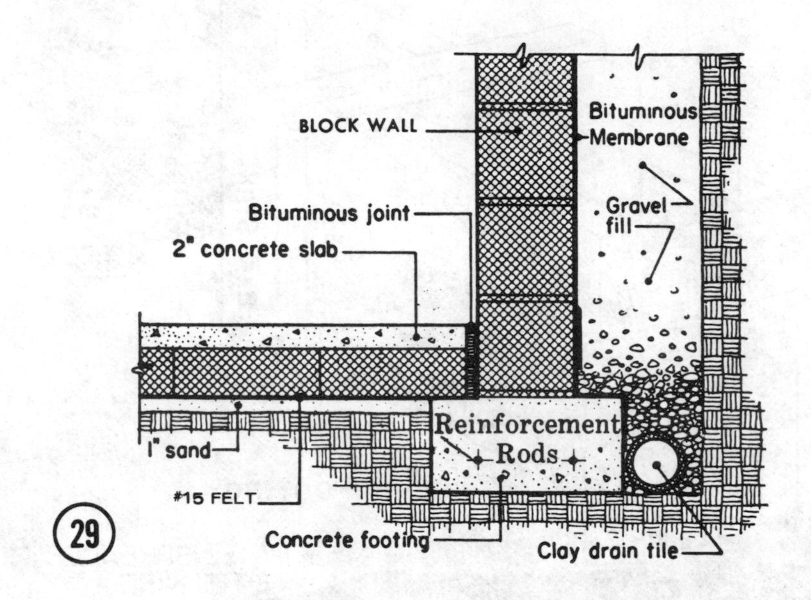

Foundation chairs, Illus. 30, simplify the positioning of reinforcing rods in footings. Space these the distance needed to keep rods level.

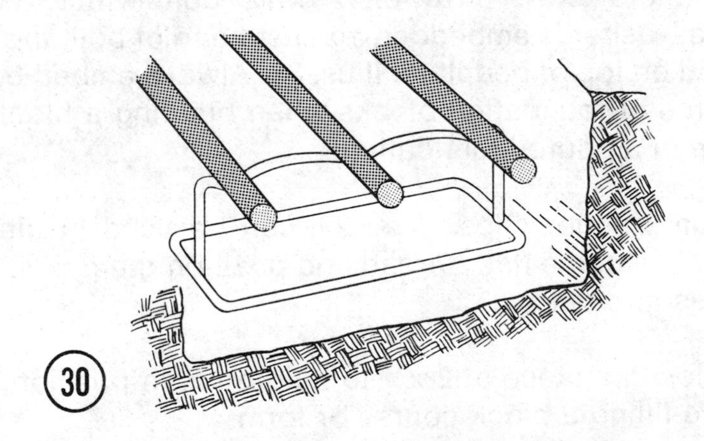

A build-it-yourself footing form consists of 2 x 6, 2 x 8, or 2 x 10 sides A, Illus. 31; stakes B, 1 x 2 spreaders C and ties D.

Use 2 x 2 or 2 x 4, sharpened at one end for stakes. Always drill holes about 1″ up from bottom of form every 6′0″. Only tack spreaders in place. Stagger ties 3′0″ from spreaders. Predrilling holes permits removing spreaders without disturbing forms. Remove 1 x 2 spreaders as you fill form with concrete.

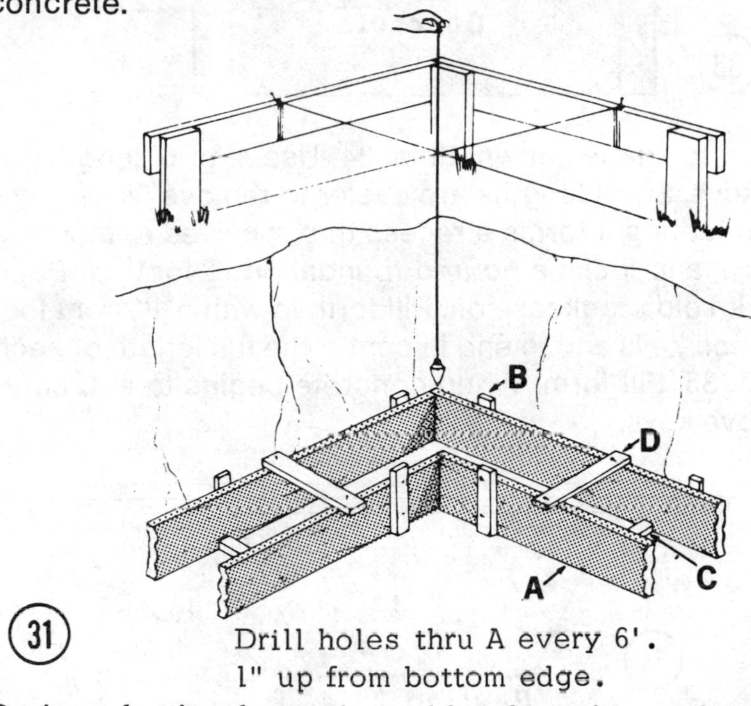

Drill holes thru A every 6'.
1" up from bottom edge.

On long footing forms, butt sides A, end to end, and nail a gusset plate, Illus. 32, to reinforce joint. Allow concrete form to set undisturbed for three days before removing forms.

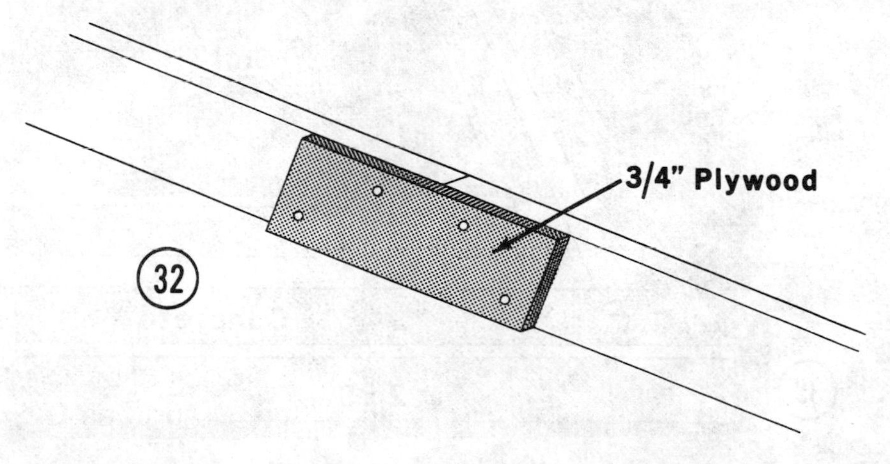

Footings for a poured concrete foundation wall, Illus. 33, should be keyed.

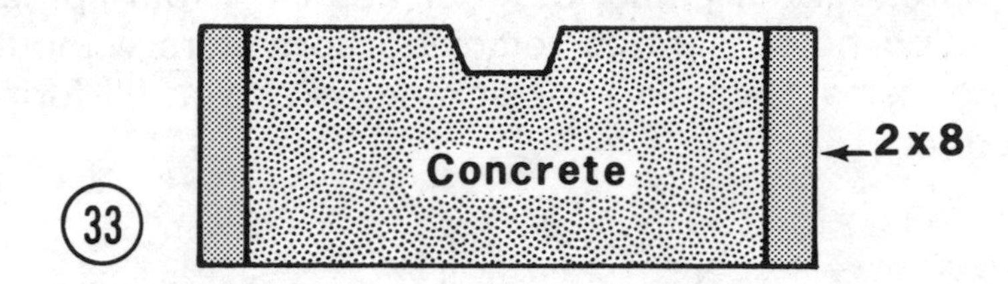

Bevel 2 x 4 to shape shown, Illus. 34. Use 4′ to 6′ lengths of 2 x 4 for keys. Short lengths are easier to remove. When embedded in footing it forms a recess that receives reinforcing rod or wire, and locks a poured foundation to footing. Paint the key with old crankcase oil. Fill form to within 1″ from top, then position keys end to end in center the full length of each form, Illus. 35. Fill form. When concrete begins to set, carefully remove keys.

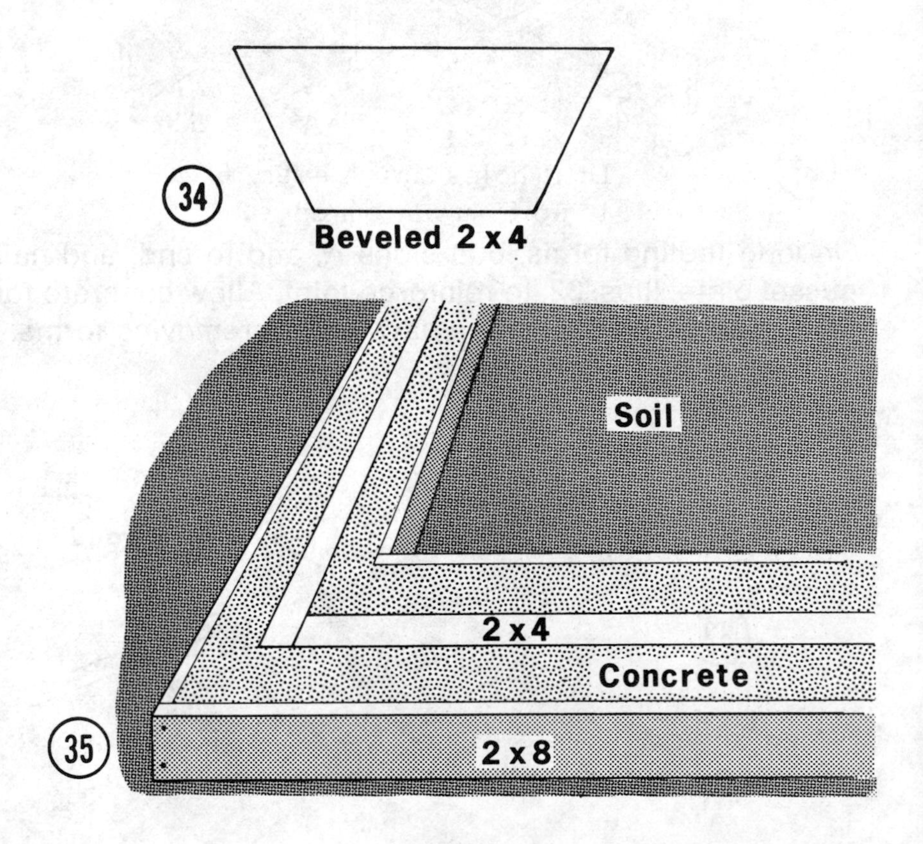

The height of a poured foundation wall and the back pressure it is subjected to, are factors that must be reckoned with when specifying size of reinforcing rod or wire. Your concrete mix dealer can recommend size and spacing of reinforcing rod.

Illus. 36, shows various size wire and rod reinforcing currently available.

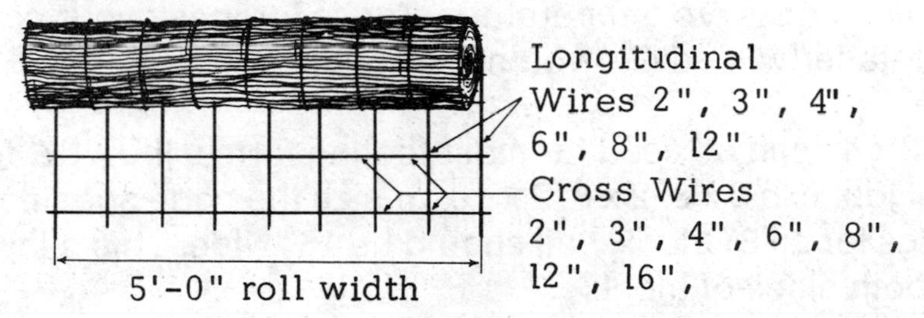

ELECTRICALLY WELDED WIRE FABRIC REINFORCING

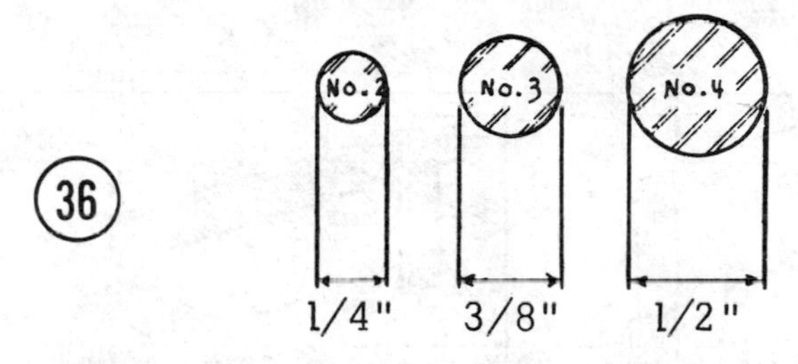

Since most first timers do work during warm weather, consider these work rules. Keep sand and water as cool as possible. Cover sand with a wet tarpaulin. Don't work during the hottest part of the day. Sprinkle the sub-grade and wood forms to cool them before pouring concrete. Sprinkle gravel before mixing. Cover the footings with burlap as soon as possible and keep the burlap wet. Keep the concrete constantly wet and, if possible, don't allow it to dry during a seven-day curing process. While many builders don't take these precautions yet still do creditable work, curing concrete properly insures better workmanship.

In cold weather, always protect the concrete from freezing. When air temperature ranges between 30° and 40°F, heat the water before mixing concrete. When ordering ready-mix, keep the trench covered with straw or other insulating material to prevent frost from entering form. Keep all reinforcing wire covered so no ice forms. Never pour concrete into a frozen form. The Portland Cement Association recommends using 1 lb. of calcium chloride to each sack of cement to hasten hardening during cold weather. Don't use calcium chloride with any other hardening or anti-freeze additive. Always cover freshly laid concrete to retain the heat. Use a tarpaulin. If a severe temperature drop is forecast, your concrete retailer will advise you not to pour.

While 2 x 4s can be used for most footing forms, Illus. 37, for a small job, use 2 x 6, 2 x 8, 2 x 10, or size the code specifies. Footings for an 8″ block wall should be 16″ wide. This allows 4″ on both sides of block.

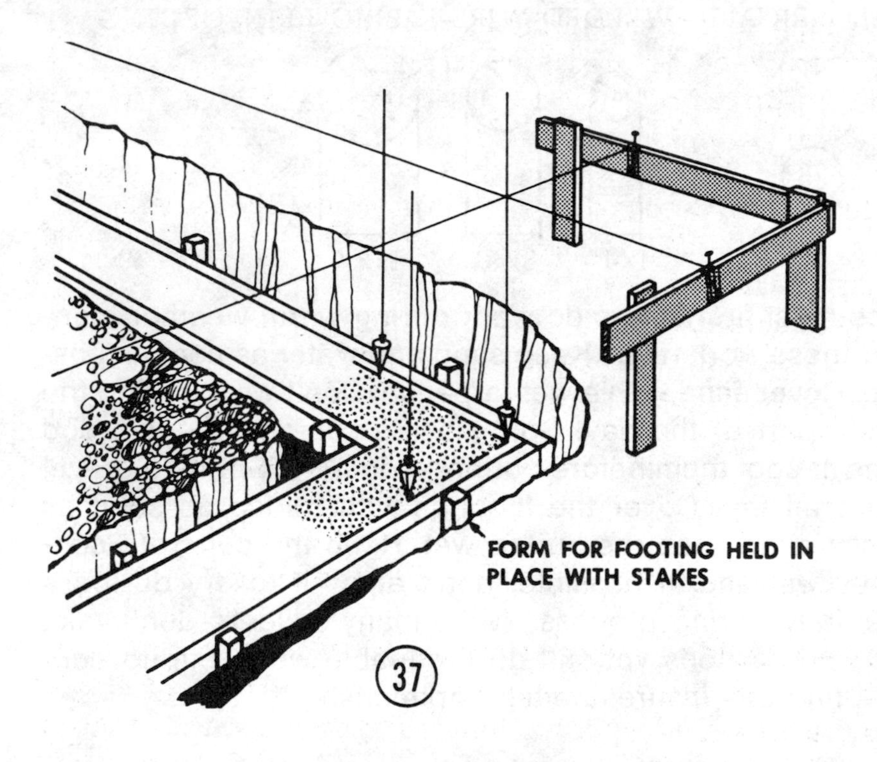

37

Footing forms are usually set up to width equal to twice the thickness of foundation wall. An 8″ x 8″ x 16″ or 18″ block wall requires a 16″ wide footing; a 10″ block or poured concrete foundation requires a 20″ wide footing. Most codes specify minimum thickness of footings.

The top edge of a footing form must be level and an exact distance from guide lines that can be divided by 8″. Estimate the number of courses you plan on laying. Mark each course off on a layout stick, Illus. 13, then add the thickness of the footing. This will be 5½″ if you use 2 x 6; 7¼″ for a 2 x 8; 9¼″ for 2 x 10. Use the layout stick to make certain that footing trenches are the same distance from guide lines. After trenches have been excavated to depth below frost level that local codes suggest, cut off length allowed for footings and use the pole to test distance from top edge of footing form to guide line all around.

When digging a footing trench, it's important to dig only to depth required and not throw back loose soil to fill in. Use stone or gravel to fill in a low spot. Use the plumb bob to guide you in establishing exact corners.

Use lumber for footing forms edgewise, Illus. 35. Check same with a carpenter's level. Lumber selected for forms should be straight and have a straight edge. Hold the footing forms in position by driving stakes into ground. Toenail form to stakes. Use 1 x 2 spreaders to hold forms apart to width required and 1 x 2 ties to hold them together.

When laying a deep footing form, it's perfectly all right to use clean fieldstone to conserve concrete. Be sure there's no dirt on the stone. Wet the stone before filling with concrete. Use a stick to poke concrete in and around fieldstone.

If you are building a footing form against an outcropping of rock, Illus. 38, build the form up to the rock, then step it up over the rock. Embed horizontal reinforcing rods, as shown, Illus. 29, but use reinforcing of sufficient length to reach to end of footing. The rods tie the footing together and to the rock.

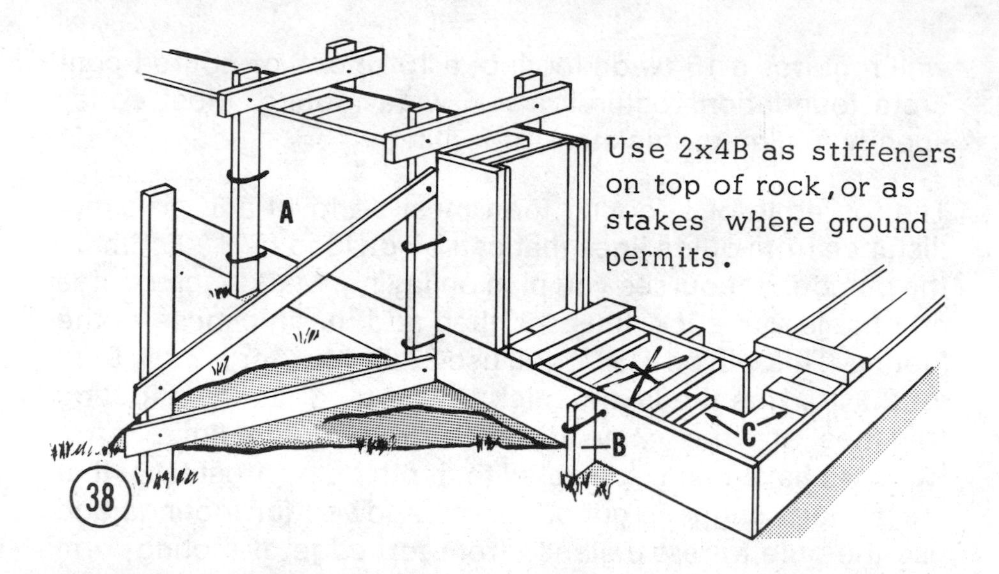

FOUNDATION FORM

Use ties, Illus. 39, or wire, Illus. 40, to hold forms together. 16-gauge galvanized tie wire can be used as shown. Drill holes and run wire around 2 x 4 stakes B, Illus. 38.

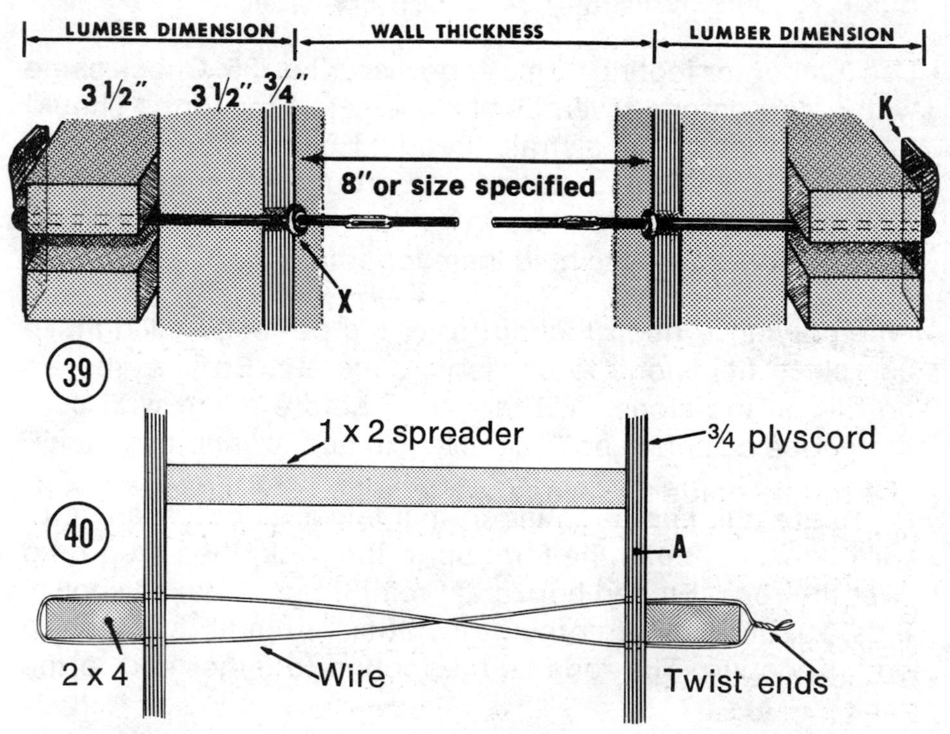

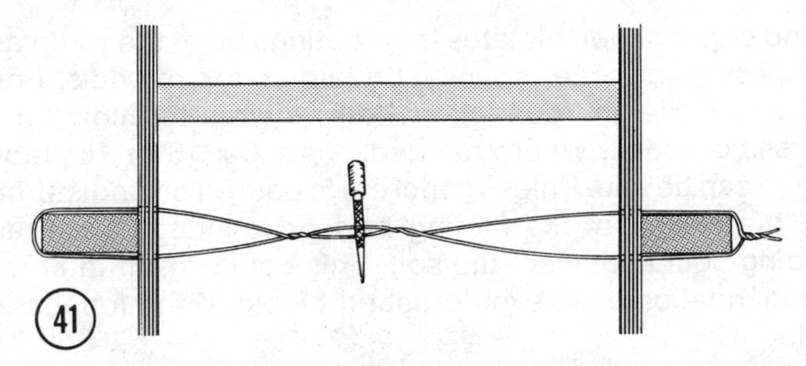

Twist wire tightly around stake or stud. Insert a 1 x 2 spreader 1″ or 2″ below or to one side of wire and twist the wire taut, using a nailset, Illus. 41.

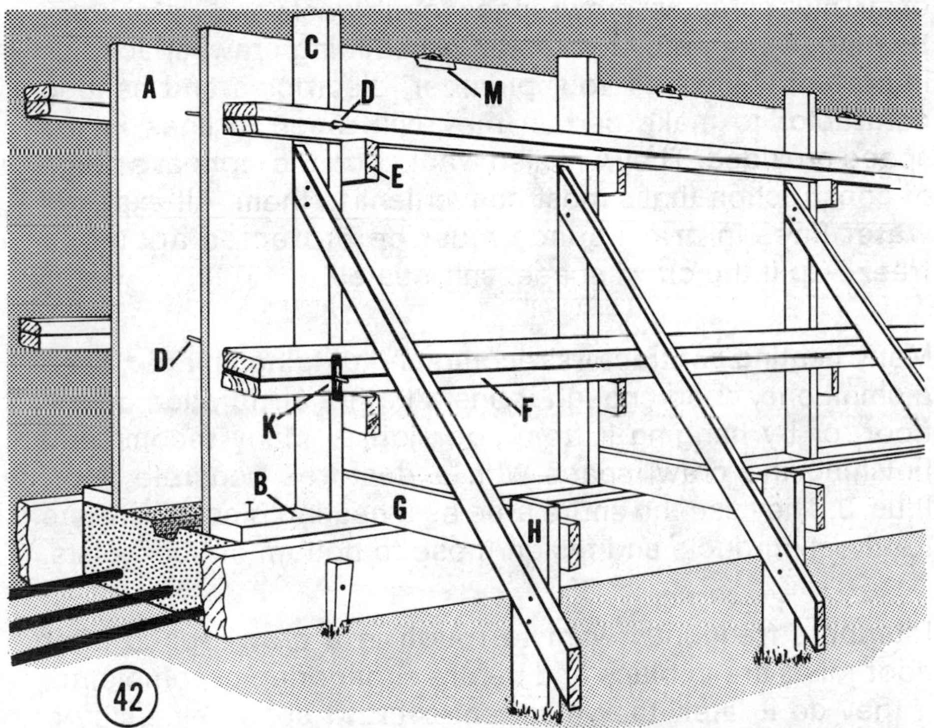

The rod tie holds forms apart at X, Illus. 39, while wedges K, Illus. 42, driven into stops on ends of ties, hold forms together.

A— ¾ plyscord
B— 2 x 4 or 2 x 6
C— 2 x 4
D— rod or wire ties
E— 2 x 4 ledger
F— 2 x 4 walers
G— 2 x 4 bracing
H— 2 x 4 stakes
K— wedge
M— channel ties

Sound construction dictates that footings be of the width and thickness local codes specify, be laid on top of undisturbed soil, and to height foundation wall requires to allow for 8″ courses of block. Where needed, a 4 x 8 x 16 or 18″ block course can be laid. Unless concrete is poured on undisturbed soil, the weight of the footing and/or foundation wall and building could compact the soil. This could result in cracks in a foundation wall, even crushed blocks if the foundation settled.

If you don't plan on building a full basement, excavate at least 48″ beneath floor framing to provide a free circulation of air, plus sufficient crawl space to run waste and water lines, heating, wiring, etc. If you plan on building crawl space, it's important to consult your plumber, electrician and heating contractor to make certain they will rough in lines in the space provided. They will also want to do the work at a stage of construction that's most convenient to them. All exposed water lines in crawl space must be protected against a freeze-up if the crawl space isn't heated.

Many heating contractors recommend digging an area, even a small one, deep enough to install a hot air furnace on the floor, or by hanging it from floor joists. Many recommend finishing the crawl space with a dust-free concrete floor, Illus. 3. They use the entire area as a heating plenum. Others run heating ducts and fasten these to bottom of floor joists.

Plumbers frequently want to rough in a crawl space after floor joists are in place and before subflooring is completed. If they do it later, they must enter crawl space either by a window placed in the top three courses of block or through a trap door cut in the floor.

PREFABRICATED FOUNDATION FORMS

Building forms and pouring a steel rod or wire-reinforced concrete foundation isn't difficult and doesn't require too much labor if a readymix truck can chute the load into the forms. If your site is wooded, rocky, or so situated that you must mix the concrete on the job, or wheelbarrow it from the truck, use blocks unless you have a supply of strong backs in search of physical exercise.

Contractors use prefabricated forms when pouring foundations, Illus. 43. These save time and labor and are well worth rental costs. If a tool rental, lumber or concrete products dealer doesn't have forms available, and there are none to be had close by, consider building your own and renting these after you finish your job. Since the ¾" plyscord needed for forms can be reused for sheathing or subflooring, there's little waste even if you don't go in for renting.

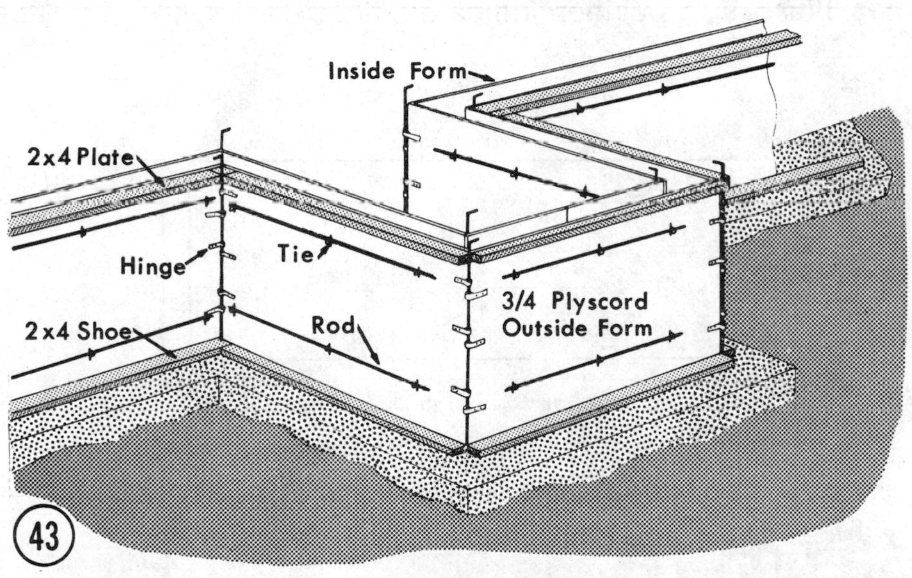

Teachers and others who need a reusable or rentable form should consider assembling this commercial type. It's easy-to-erect, easy-to-strip and with normal care can be reused endless times. Directions for making and assembling this form are provided on page 52. Ask your concrete products and tool rental dealer about rental forms. If same are not available locally, making a set might be the start of a new business.

If you plan on erecting a prefab, talk to the company that sells them. If you make a set of forms to pour your foundation, they could help other prefab buyers by renting your forms. Use plastic coated plywood if you plan on making reusable forms.

Those building a form for one time use who want to salvage and reuse all materials, should build the form shown, Illus. 42. Use ¾″ plyscord A, 2 x 4 or 2 x 6 for shoes B; 2 x 4 studs C spaced 24″ on centers, rod or wire ties D, 2 x 4 walers F. Position walers 12″ up from bottom and at 2′0″ spacing. Ledger blocks E, nailed to studs, hold walers in position until securely fastened with wedge K. 2 x 4 diagonal and horizontal braces G are nailed to both sides of H. Drill ½″ holes for rod ties, Illus. 39, in position indicated, Illus. 44.

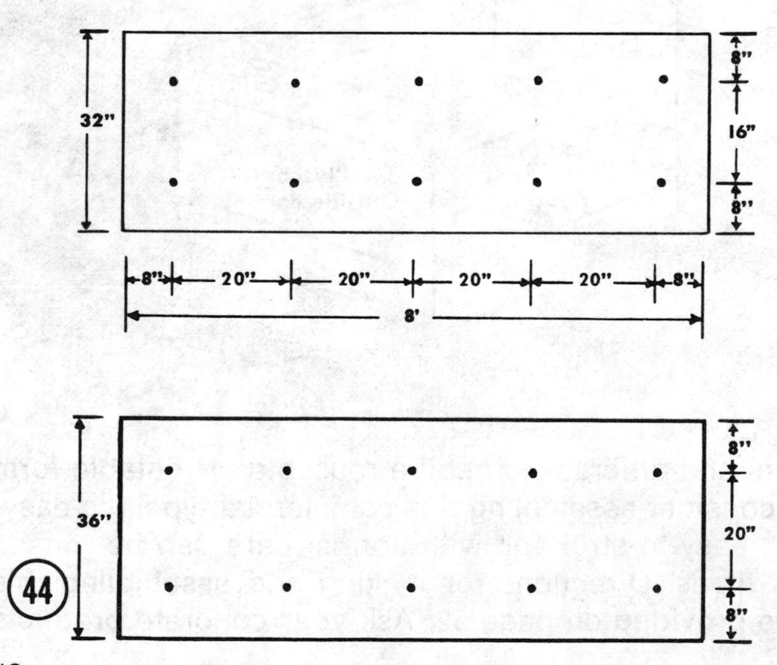

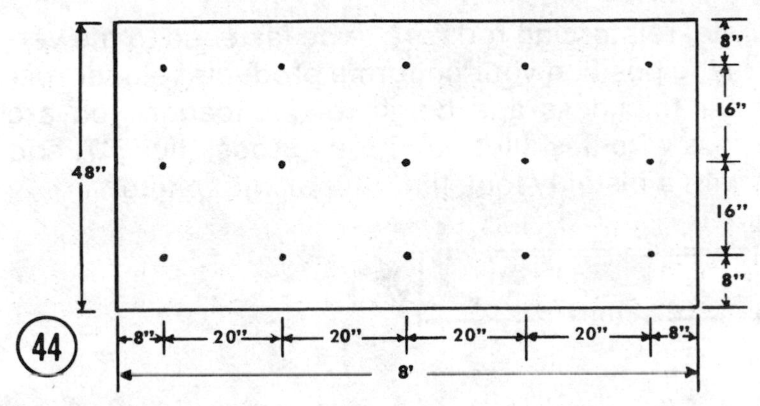

To set up foundation forms, snap a chalk line on footing 3¼″ in from edge of a 16″ wide footing. Check diagonals to make certain lines are square. Nail straight 2 x 4 shoes to "green concrete" (within 48 hours of pouring), or nail 2 x 6 to footing form in position required.

Nail sides to studs. Keep studs 1½″ up from bottom edge of side so stud rests on shoe. Drill holes and bolt or nail diagonal and horizontal 2 x 4 bracing G to studs and to 2 x 4 stakes. When inside form is plumbed and braced in position, install reinforcing rod or 6 x 6 wire in center of form, Illus. 45.

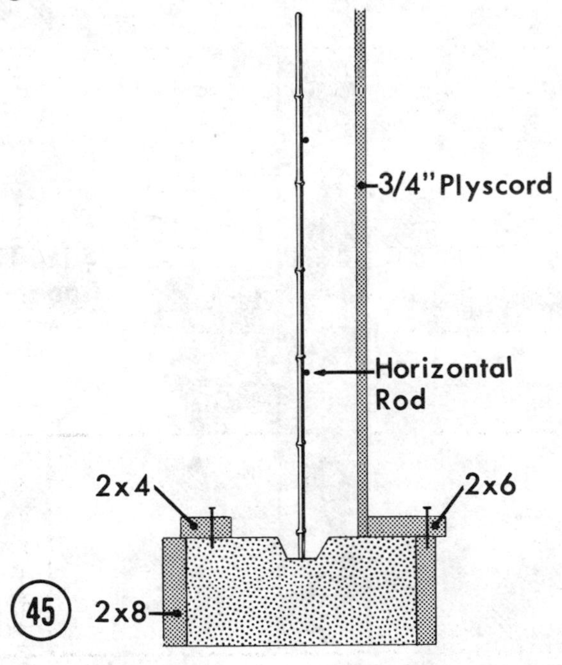

The horizontal reinforcing rods are to be fastened to the vertical rods in the position your concrete products retailer recommends for thickness and height of foundation you are pouring. Press wire ties, Illus. 46, around rods, Illus. 47, and twist tight with a pistol ty tool, Illus. 48, or with pliers.

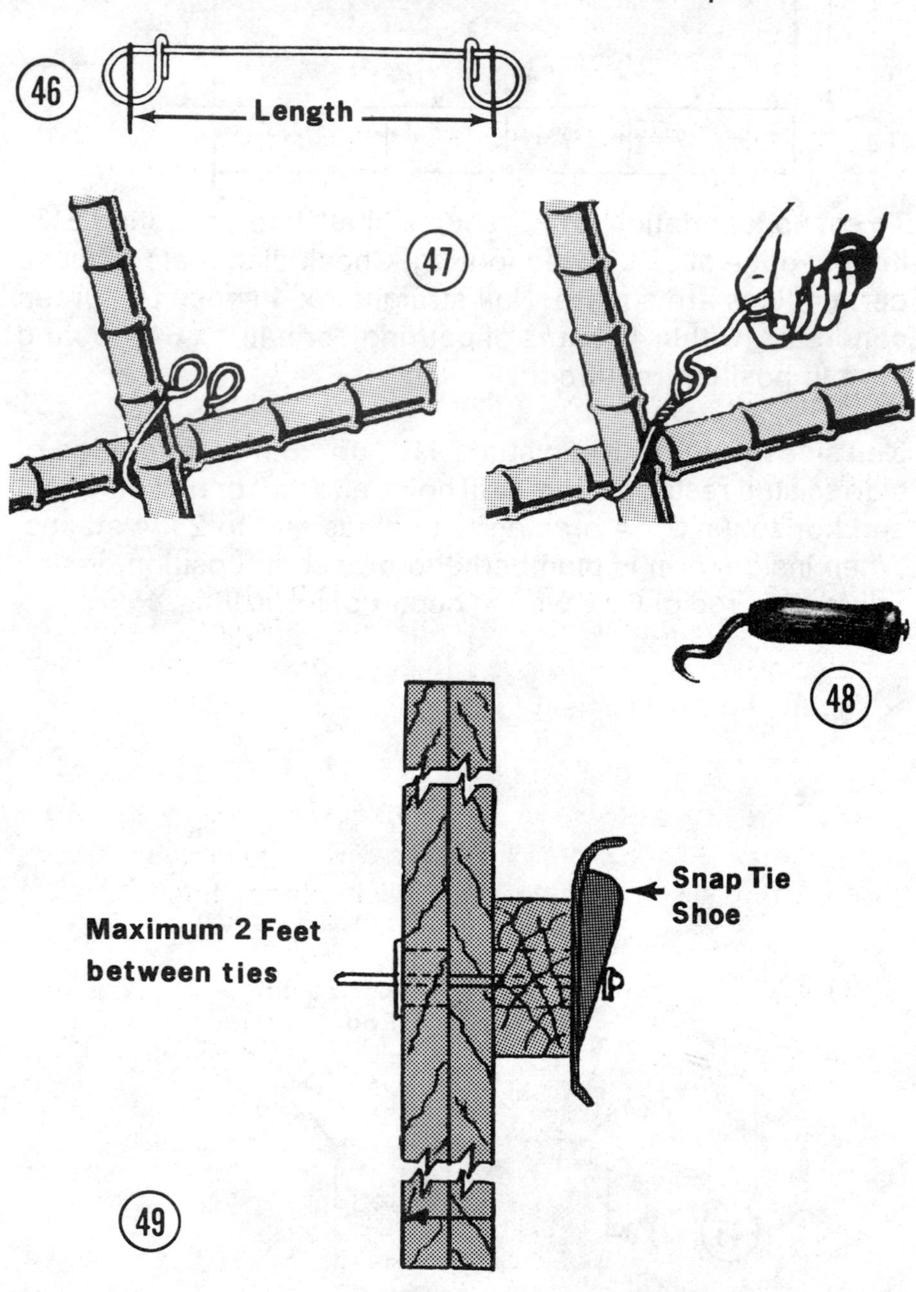

Insert rod ties, Illus. 39. Fasten reinforcing to ties in center of form. Next install outside form. Position 2 x 4 walers on ledger blocks, drive snap tie shoes on ends of ties, Illus. 49.

The rod ties hold forms apart and together. If you decide to use 16 gauge wire, use 1 x 2 spreaders. The 1 x 2 must be removed as you pour concrete.

Your concrete mix company can suggest thickness of form, spacing for studs, walers, etc., when you explain height and size of foundation, height of slope adjacent to a foundation, etc., etc.

Don't start erecting forms until you have allowed footing to set 48 to 72 hours. Don't pour a foundation or build a block wall until footing has been allowed to set three or more days.

Always order readymix a day or two in advance so you can get an early morning delivery. This permits pouring the entire job in one day. Use a 1 x 2 as a plunger on a small job; rent a vibrator to eliminate any air pockets on a big job. Allow foundation to set at least three days before stripping forms. Depending on whose ties you use, these normally break off flush or slightly below surface of wall.

Your concrete products dealer rents special ties for holding a basement window frame, Illus. 50, in place when pouring a foundation. Brace frame to inside form. Ties fastened to a 2 x 4 x 3′ or 4′, hold form and window in place, Illus. 51.

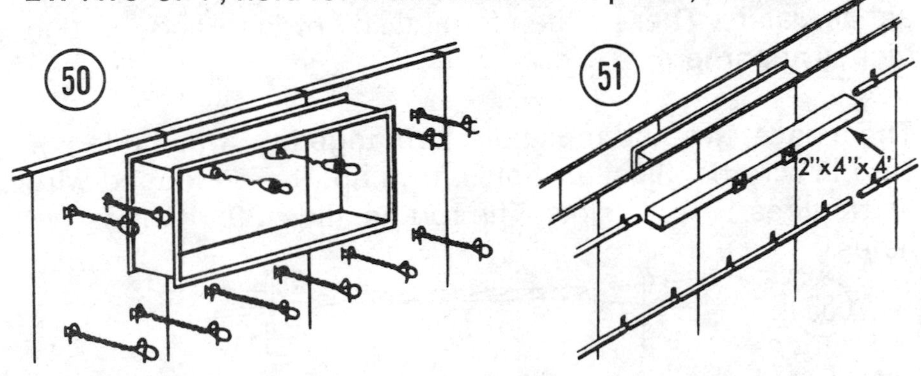

A reusable rental form, Illus. 52, consists of ¾″ plyscord sides to height foundation requires, special looped end wire ties, Illus. 53, hinges, cane rods for hinges, plus diagonal bracing nailed to 2 x 4 plates.

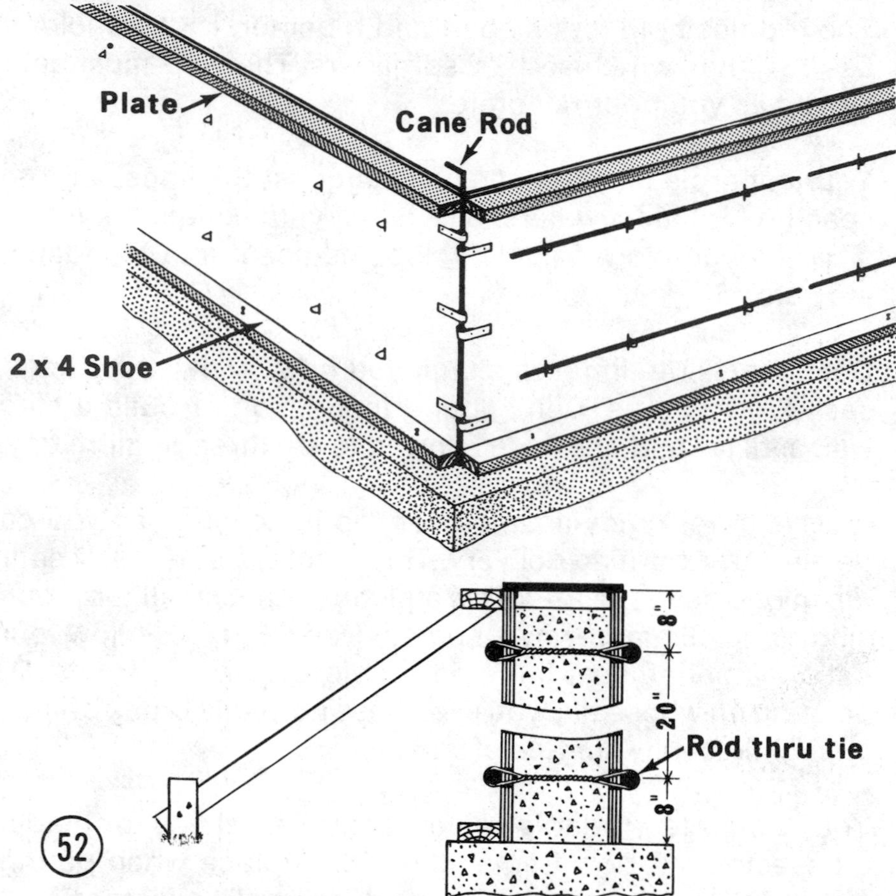

Rods placed horizontally through ends of wire ties, Illus. 43, act as walers. These stiffen form. Cane rods thread through hinges at corners.

The hinges, ties and rods hold form together, while a leg or stop in wire tie, Illus. 53, holds form apart. The looped wire tie requires ¼″ x 1″ slots. The rod tie, Illus. 39, requires ½″ holes.

53

For a 32″, 36″, or 48″ high foundation, position ties as shown, Illus. 44.

To build reusable forms, saw or plane ends of corner panels 45° to shape shown, Illus. 54. ¼″ holes drilled in position hinge requires permits bolting hinge straps, Illus. 55.

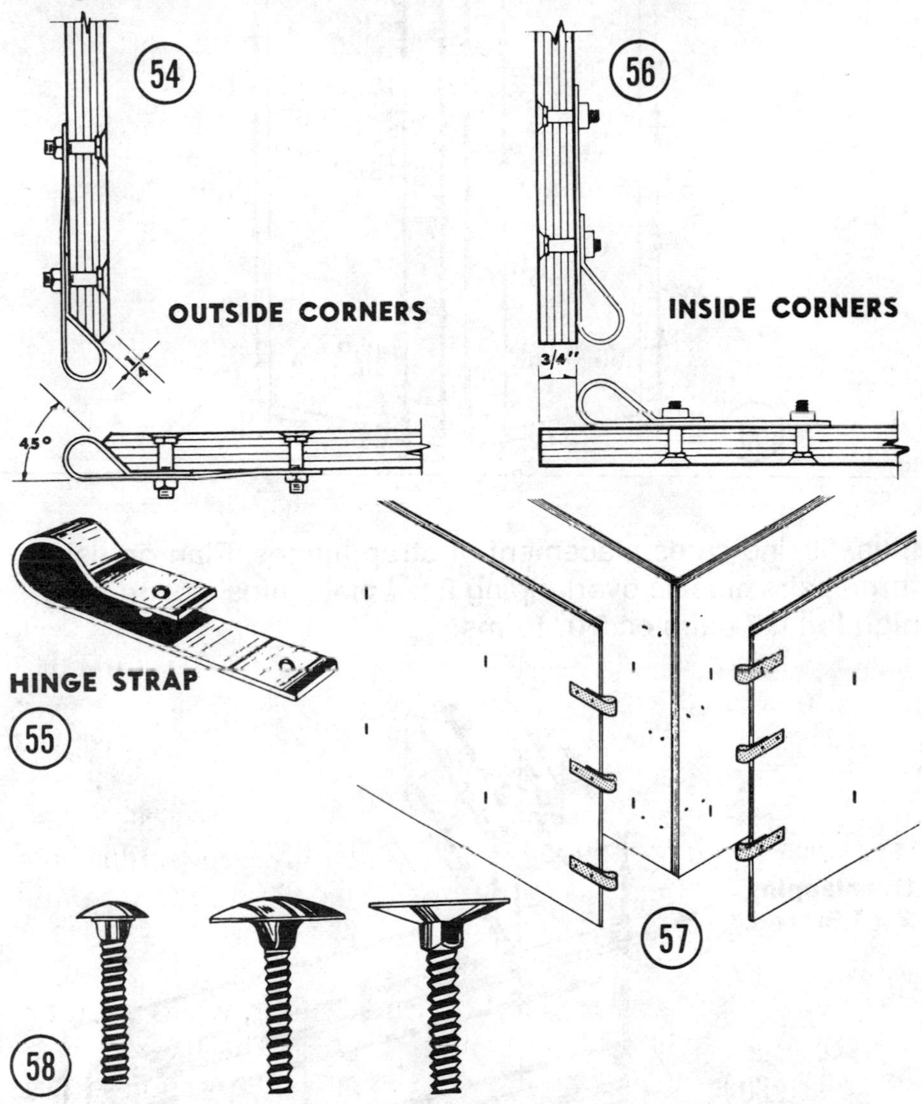

Hinge straps fastened in position to inside form, Illus. 56, 57, permit locking forms at corners with cane rods supplied by hinge manufacturer. Use ¼ x 1¼″ carriage bolts, Illus. 58.

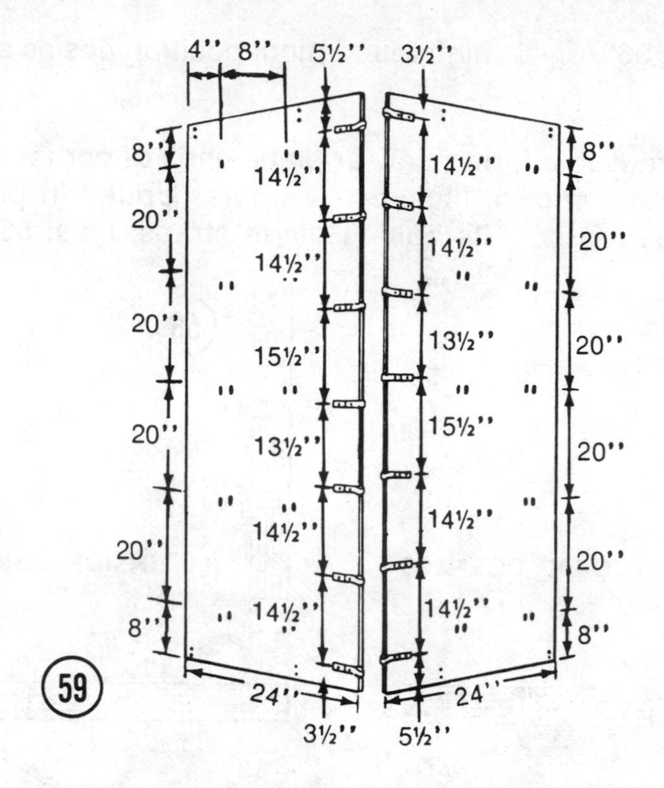

Illus. 59, indicates placement of strap hinges. Plan on using three pairs plus an overlapping 2 x 4 plate, Illus. 60, on a 4′0″ high form; 7 pairs on 8′0″ forms.

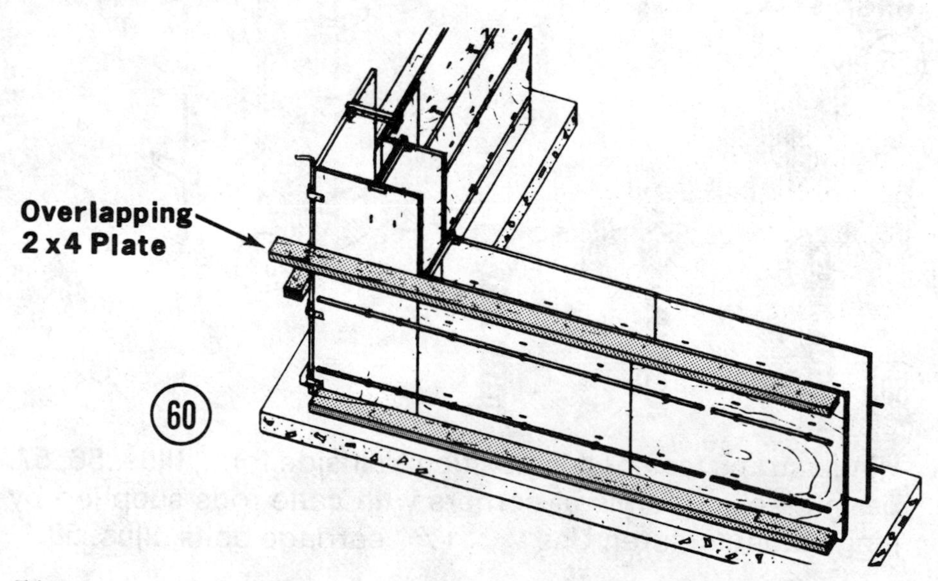

The hinge straps, rod canes, rods, channel tops and looped wire ties are currently available from suppliers of concrete form hardware. Check the Yellow Pages. Buy ¾″ plastic coated plywood for top quality reusable forms.

If you plan on making any quantity of these forms, cut a ⅜″ x 3′ x 8′ plyscord template, Illus. 61. Using a sabre saw, make 6″ x 7″ cutouts in position noted.

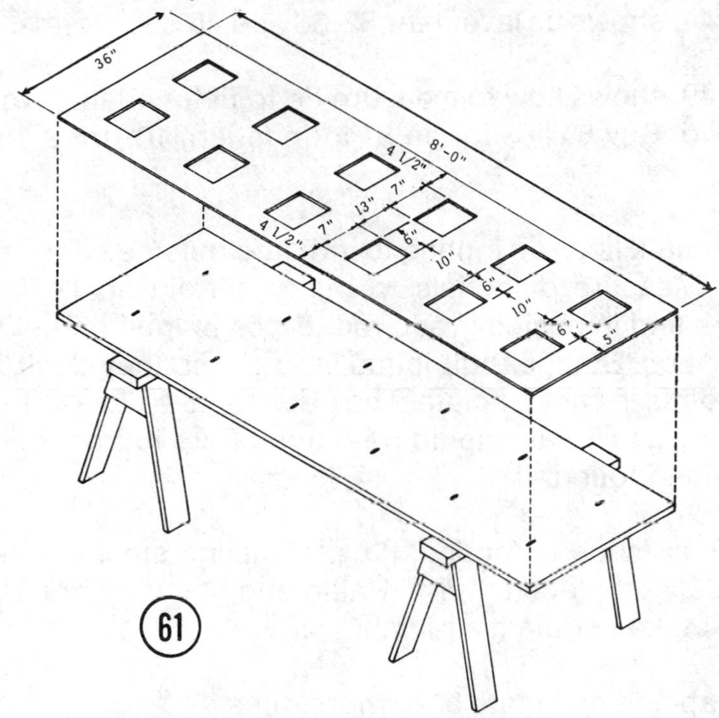

To simplify routing ¼″ x 1″ slots in position, cut a ⅜″ x 6″ x 6″ block, Illus. 62. Drill a ¼″ hole in center. Insert bit, Illus. 63, in router, bit through block. Clamp ⅜″ x 3′ x 8′ template, Illus. 61, to ¾″ plyscord panel and rout ¼″ x 1″ slots in position shown, Illus. 64.

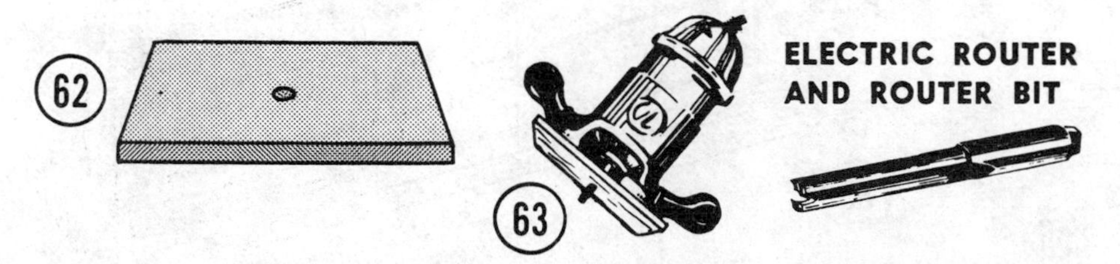

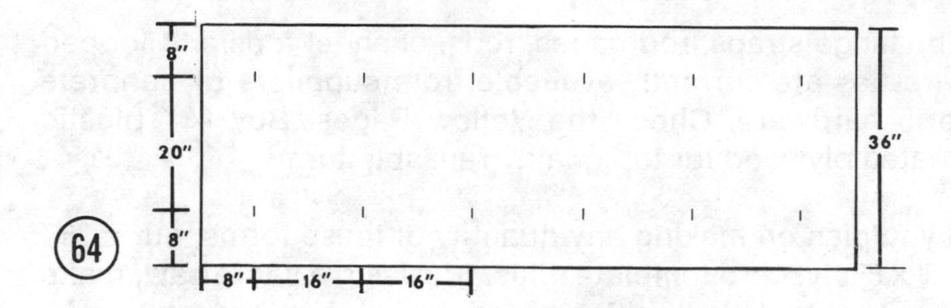

Illus. 44, shows tie layout for 32, 36 and 48 x 96″ forms.

Illus. 39, shows how to measure tie to fit foundation you want to build. Buy 8″ ties for an 8″ foundation, 10″ for a 10″ wall, etc.

Snap chalk lines on footing to indicate outside edge of inside form. Measure diagonals to make certain chalk lines are square and in position required. Place hinged corner panels in position. Stagger butt joints in forms so they don't line up with joints in shoe or plate. The outside face of a ¾″ plywood form for an 8″ wall should be placed 3¼″ in from edge of a 16″ wide footing.

Nail 2 x 4 shoe to footing,* Illus. 45, using steel cut nails, or nail a 2 x 6 to footing form. Raise and securely brace inside form. Make certain it's plumb.

Overlap and nail plate at corners, Illus. 65.

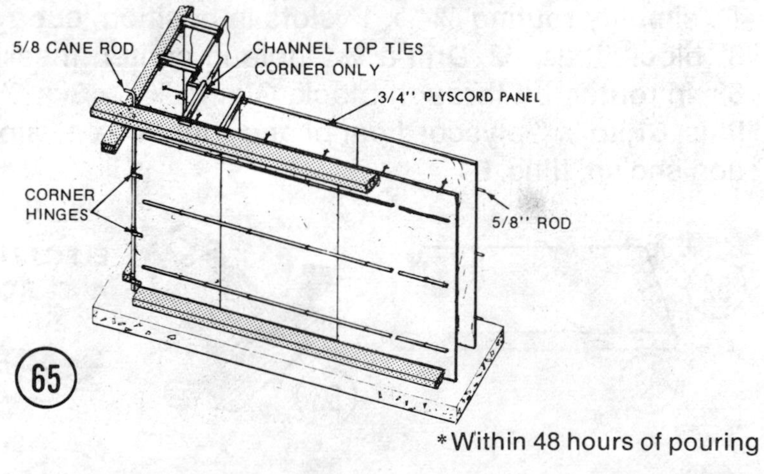

*Within 48 hours of pouring

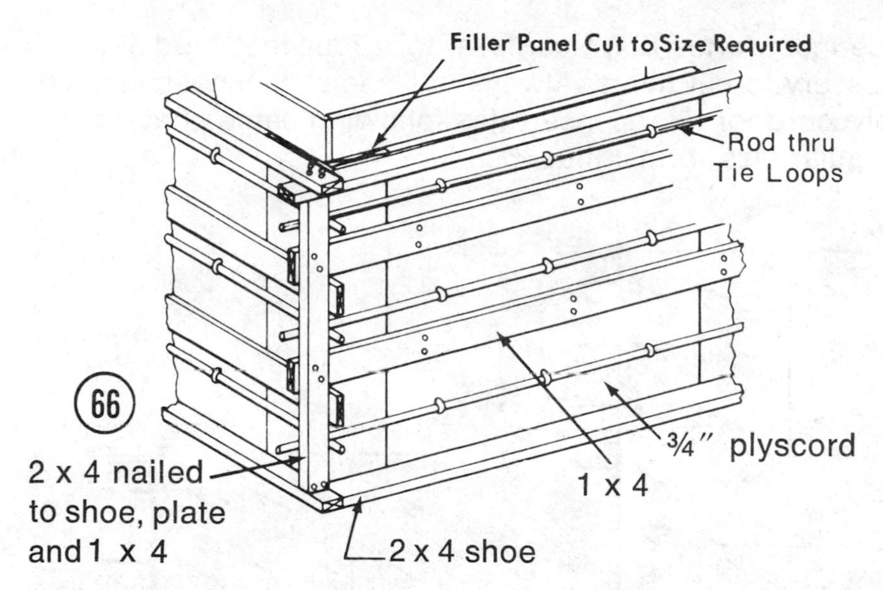

To build a heavy duty form add 1x4 or 2x4 stiffeners, Illus. 66.

Channel top ties, Illus. 67, 68, help stiffen and separate forms and can be used where it isn't possible to use ties in the top row. These come in handy at corners and when setting windows.

CHANNEL TOP TIES...

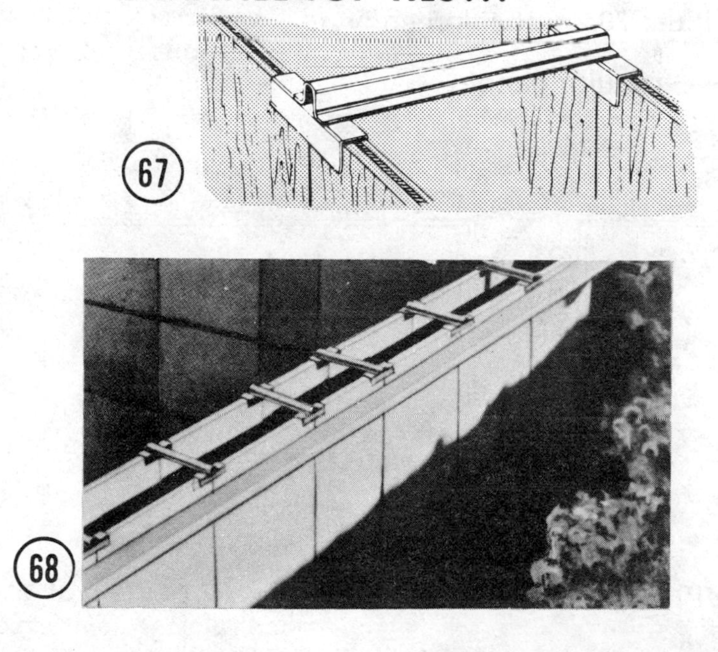

Since most forms don't conform to a four-foot module, it's necessary to cut fillers. Always use same thickness lumber or plyscord for fillers. Insert ties following same spacing as in regular size forms, Illus. 69.

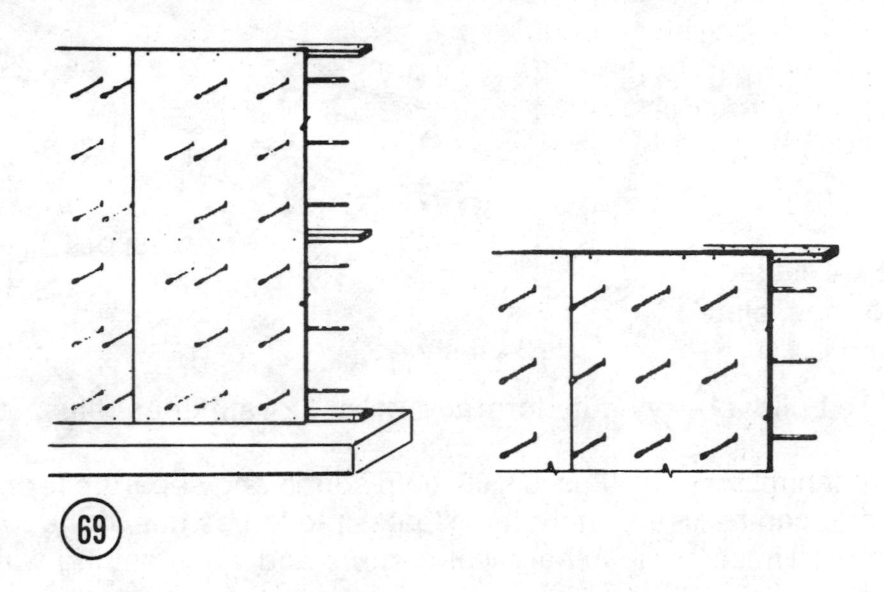

When erecting forms for a T-shaped foundation, use full panels, Illus. 70, in position shown.

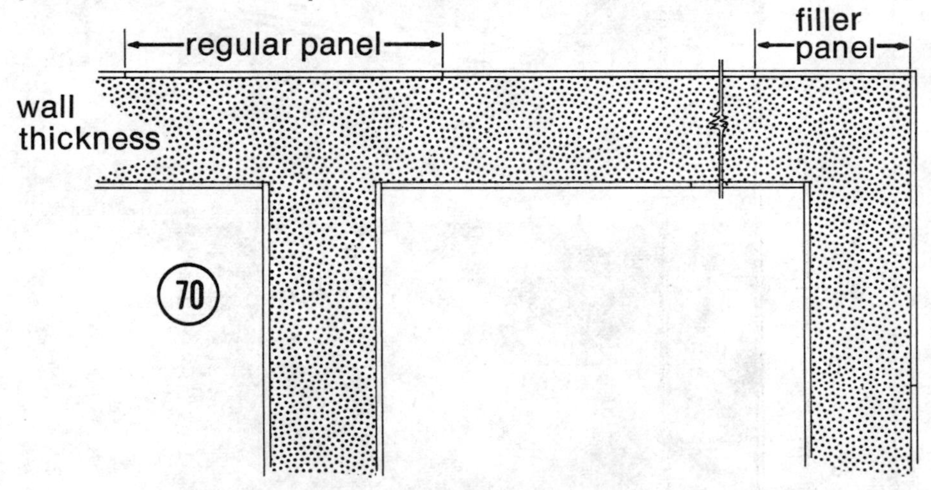

If a T form needs a filler panel, Illus. 69, 70, place it at an end.

To strip commercial forms, remove rods and corner canes. Use wire cutters, Illus. 71, to snip ties. Cut ties again flush with face of wall.

To form a ledge for brick facing, Illus. 72, nail 2 x 4's to front form in position required for bricks. Position ties so they project into brick mortar joints. When form is stripped, the projecting ties lock mortar joints to foundation. Channel top ties can be used in top course.

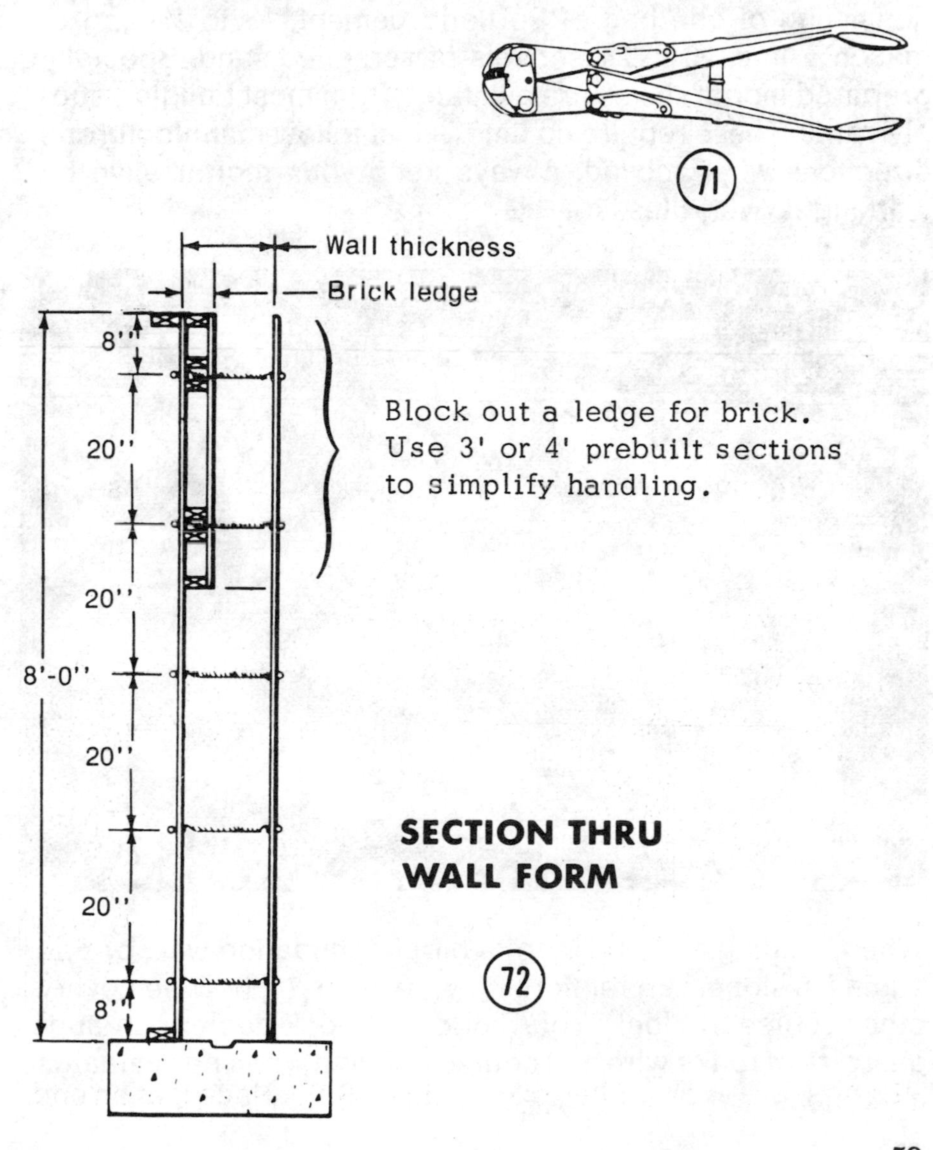

SECTION THRU WALL FORM

CONCRETE BLOCK FOUNDATION

If you decide to build a block foundation rather than pour one, use size blocks specified by local codes. Before laying blocks check guide lines to make certain they are square, level and taut. Check diagonals to make sure lines are square. Setting blocks takes some skill, common sense and a lot of practice. You can learn much by working alongside an experienced mason or by watching one work. Most masons use a mortar consisting of one bag of Portland cement, ⅓ to ½ bag of mason's lime, 20 to 24 shovels of screened sand. Specially prepared mortar mixes are available from most building supply yards. These require no lime — just follow manufacturer's directions when mixing. Always keep your mortar alive by working it over, Illus. 73.

When building a high concrete block foundation wall, or one against a slope, use reinforcing wire, Illus. 74, between every other course of block. Your concrete block dealer can suggest size of rod or wire and course spacing when he evaluates conditions.

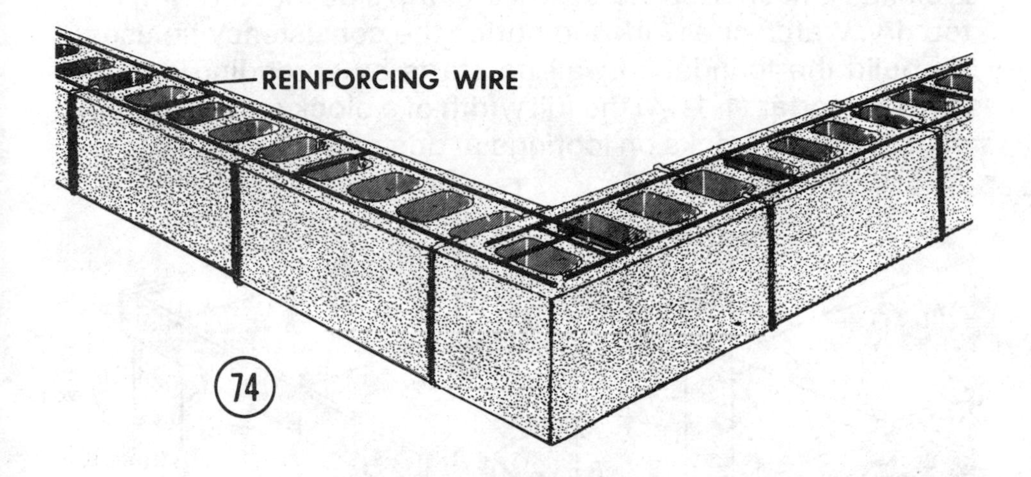

Since blocks are normally laid over a period of days, always check guide lines at the beginning of each day's work to make certain all are level and square.

Always purchase extra blocks to allow for breakage. When estimating the amount of blocks needed, figure three 8 x 8 x 16 blocks per course for each four feet; four 10 x 8 x 18″ blocks per course for every six feet. When necessary make heavier mortar joints between blocks to fill space available. Always lay end blocks, Illus. 75, on exposed outside corners above grade.

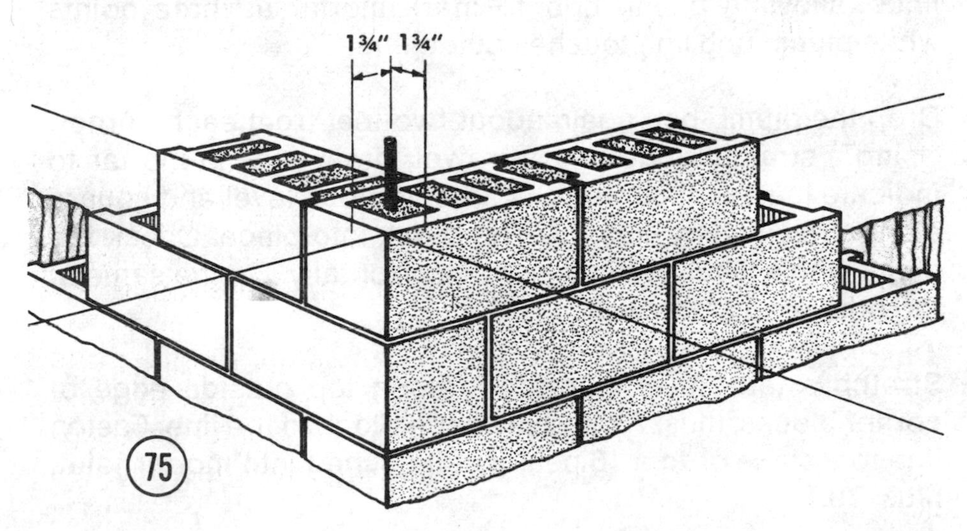

A skilled mason uses mortar on the firm side, not too wet, not too dry. Watch one work and notice the consistency he uses. To build the foundation wall he starts by spreading a thick layer of mortar (1-1½″) the full width of a block and the length of at least two blocks on footings at one corner, Illus. 76.

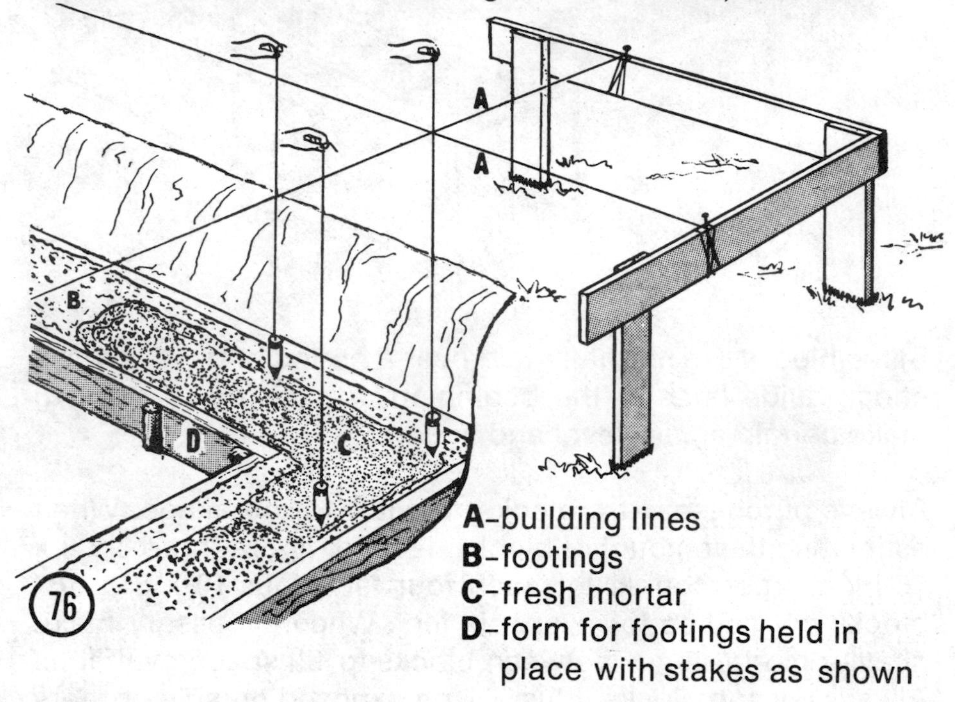

A-building lines
B-footings
C-fresh mortar
D-form for footings held in place with stakes as shown

To locate exact corner, drop a plumb bob down from guide lines, allowing plumb bob to mark mortar at three points while plumb bob line touches guide line.

Drop the plumb bob again about two feet from each corner. Using a straight edge and a trowel, draw lines in mortar to indicate the exact corner. Set an end block level and square along drawn lines, Illus. 77. Tap block into place. Check the block with a level, vertically and horizontally. Do the same at the other end of the wall.

Stretch a taut, level guide line along top outside edge of corner blocks, Illus. 78, and lay each block to the line. Fasten line to a stake or to a 16 penny nail pushed into mortar joint, Illus. 79.

77

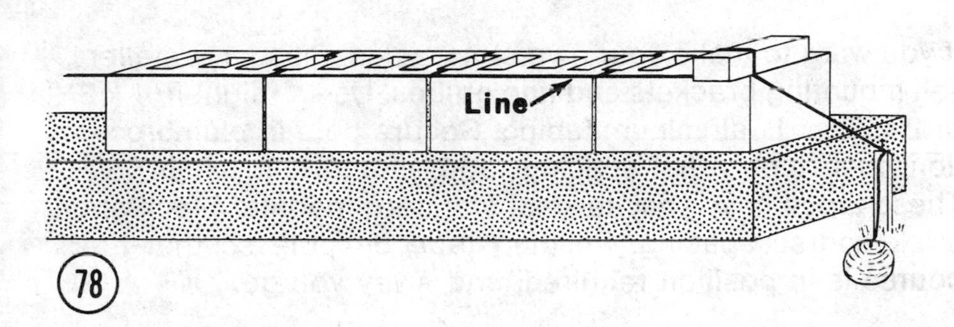

78

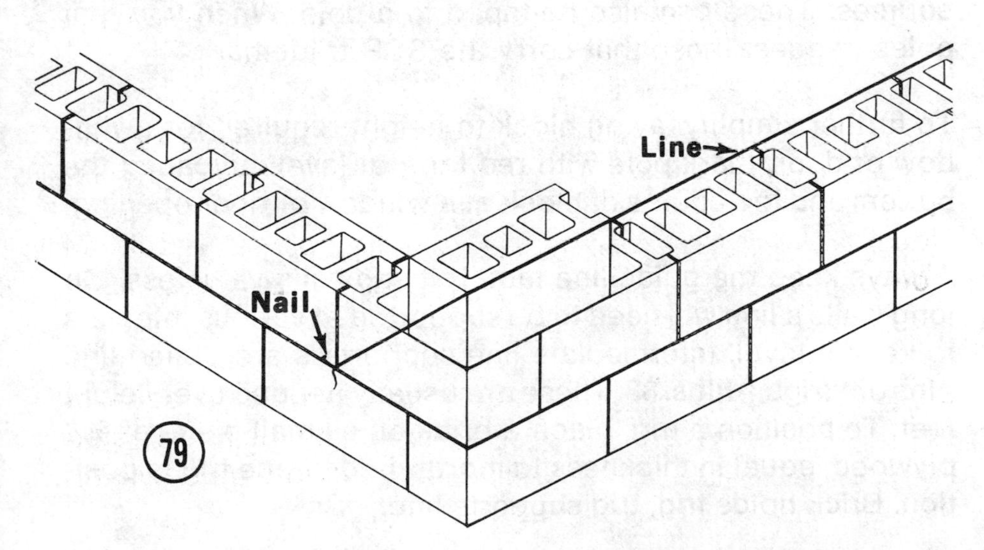

79

Laying block requires level guide lines. While you can stretch a short line between end blocks by wrapping it around a brick, the story poles, Illus. 15, spaced where needed to maintain a taut level line, help amateurs make like pros. Always check the line with a line level.

A pair of line holders, Illus. 80, sell for under $2.00. The line wraps around stem, goes through slot in fork. Tension on line keeps holders in place.

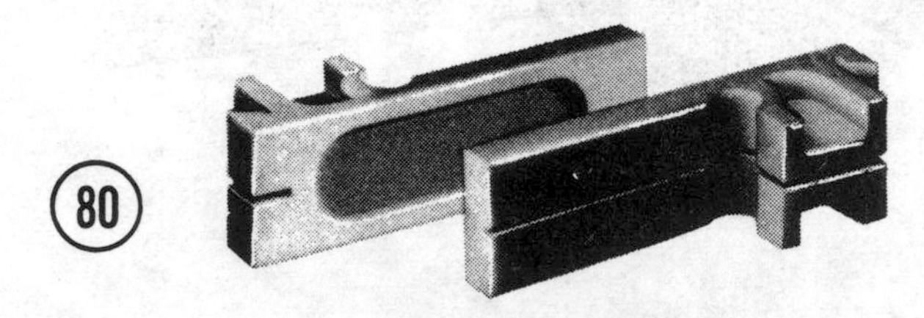

If you want to make your own story poles, masonry retailers sell mounting brackets and line guides. Use straight 5/4 x 2, or 2 x 2, or 1″ aluminum tubing. Secure pole in plumb position, then tape a flat spacing tape, Illus, 81, to each pole. These tapes have easy-to-read standard modular block and brick course spacing. Position tape on pole so that first course is in position required, and away you go.

Many folding rules, Illus. 82, also have brick and block courses. These can also be taped to a pole. When you rent poles, request those that carry the SCR trademark.

To further simplify laying block to height required for a window or door, mark pole with red tape at level indicating the bottom and top course of block at a window or door opening.

Always keep the guide line taut. If it sags, it's valueless. On long walls a line will need to be supported at various intervals to keep it level. Intermediate line supporters are called line pins or "trigs," Illus. 83. These are usually needed every eight feet. To position a trig, place a brick on a small piece of 3/8″ plywood, equal in thickness to mortar bed. Place trig in position. Brick holds trig, trig supports line.

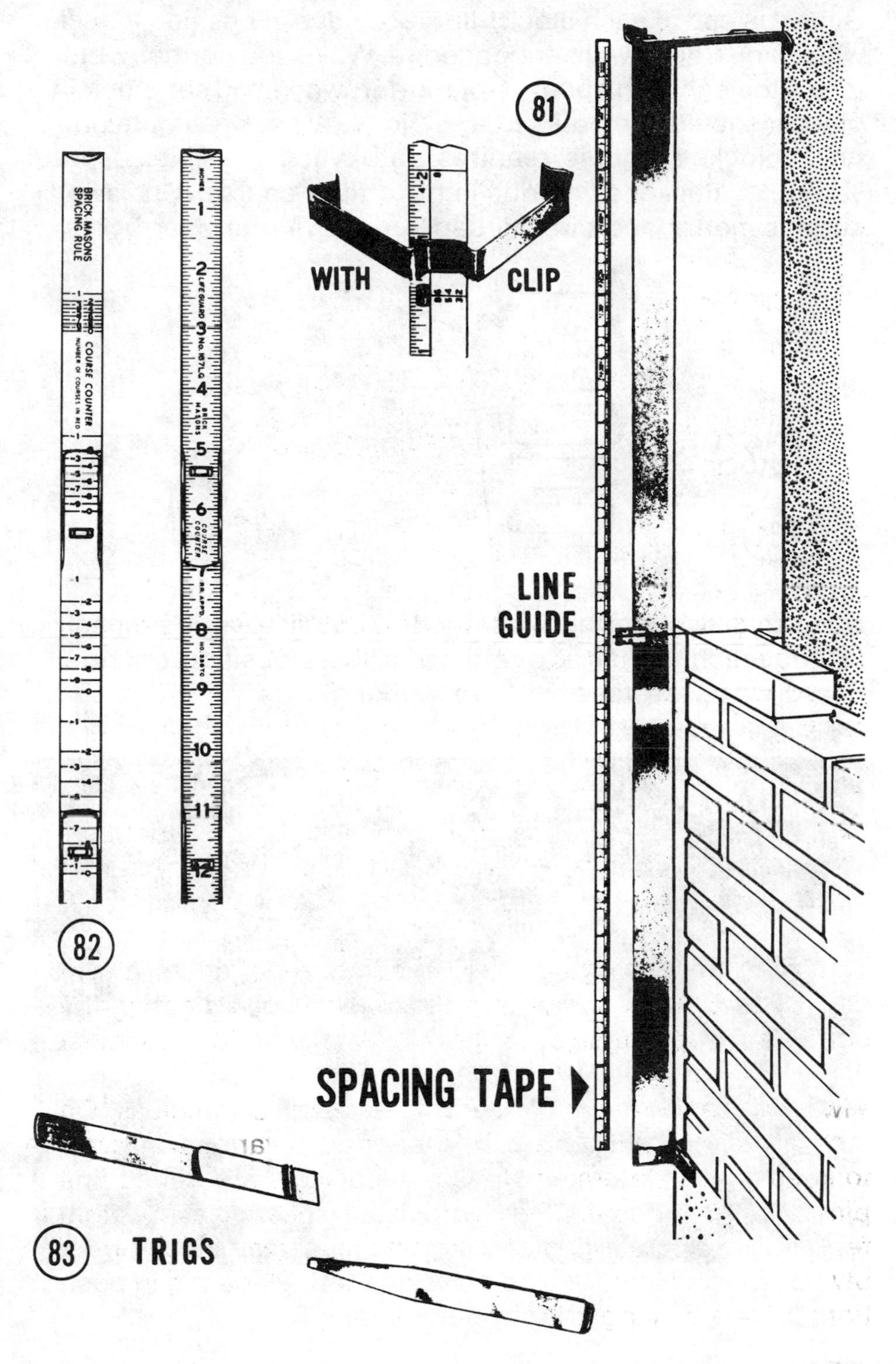
81
WITH
CLIP
BRICK MASONS SPACING RULE
COURSE COUNTER
LINE
GUIDE
SPACING TAPE
82
83
TRIGS

Butter up end of each block, Illus. 84. Place in position, check with level. Check with straight edge. While you normally butter up the end with about 1″ of mortar, when you set block in place it should compress to a ⅜″ joint. Always estimate how many blocks a course requires and, when possible, use a ⅜″ to ¾″ thickness mortar joint to fill a course. Cut away surplus mortar and throw it back on the mortar board.

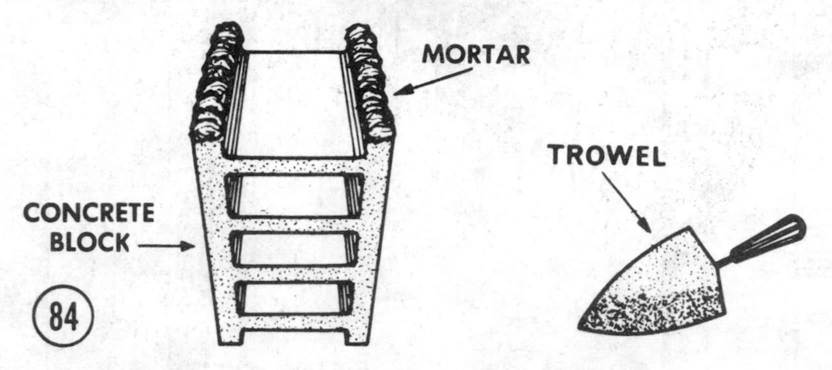

Keep remixing mortar, Illus. 85, to keep it alive. Never mix up too much mortar at one time. Always position a mortar board close to where you are working.

85

Spread mortar over the edge and cores of two or three blocks. Place, press and tap each block into position level with the line and with adjacent blocks on the same course, Illus. 86.

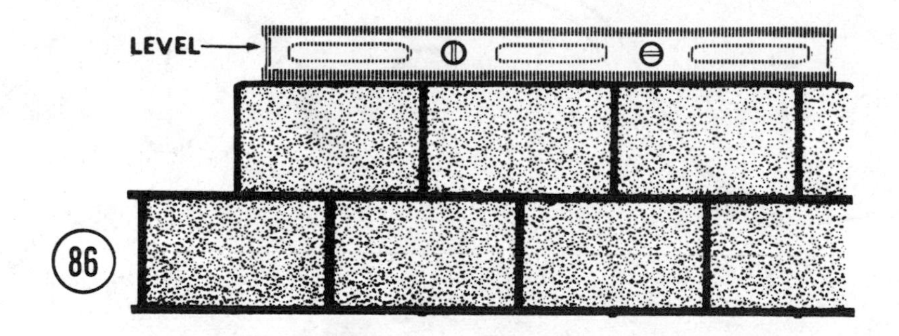

Plumb blocks vertically, Illus. 87. When setting block on each course, it is necessary to plumb the block vertically in two directions, the end and side, as well as horizontally and to the guide line. You quickly see how thick a mortar bed is required. Always cut away mortar that squeezes out of a joint. Throw it back and work it into the mortar on board.

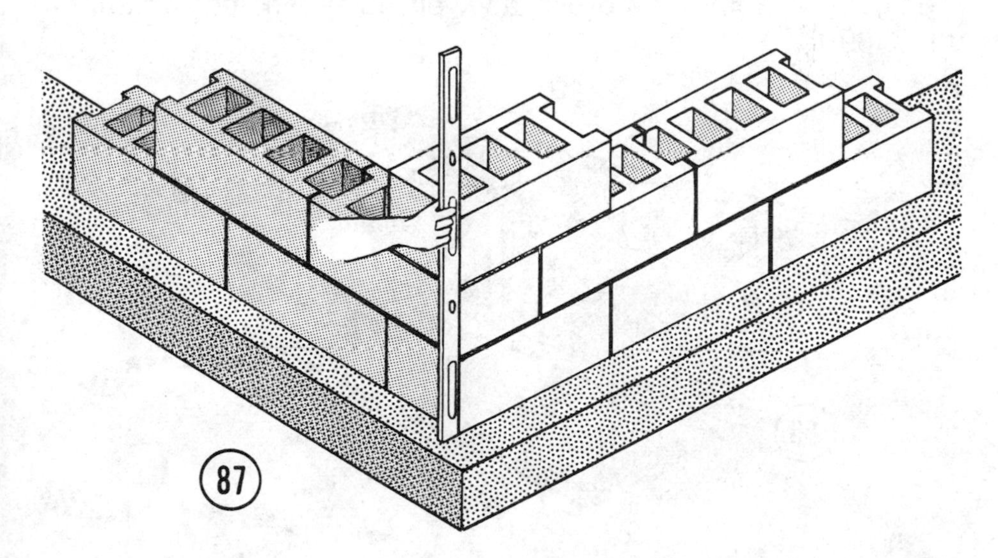

After laying a number of blocks, and before mortar begins to set, use a jointer, Illus. 88, to finish the joint. These come in various shapes, concave or V-joint.

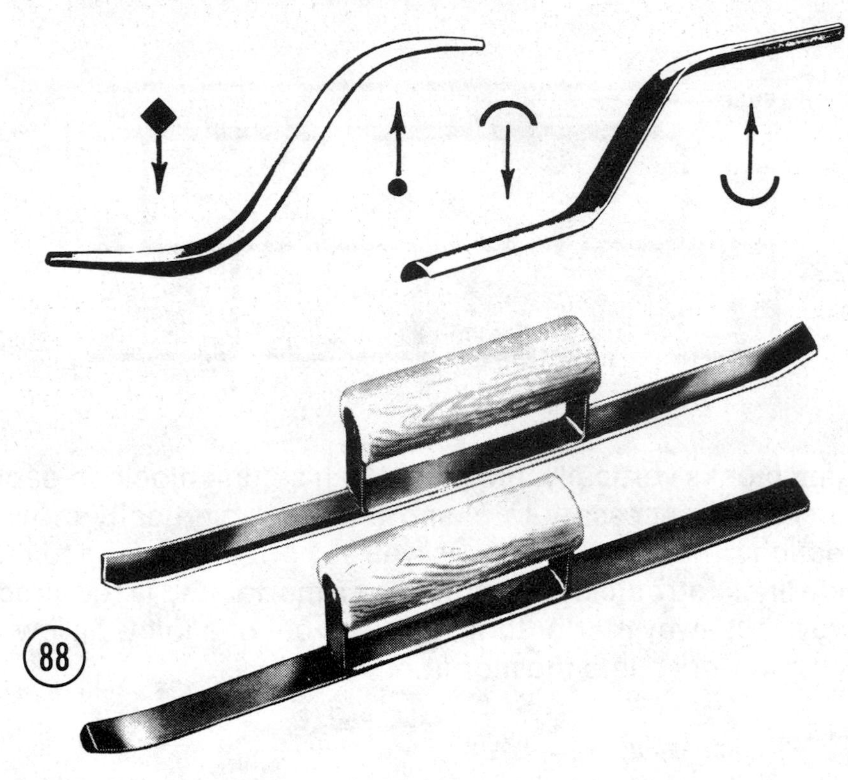

The jointer helps compress as well as finish mortar joint, Illus. 89.

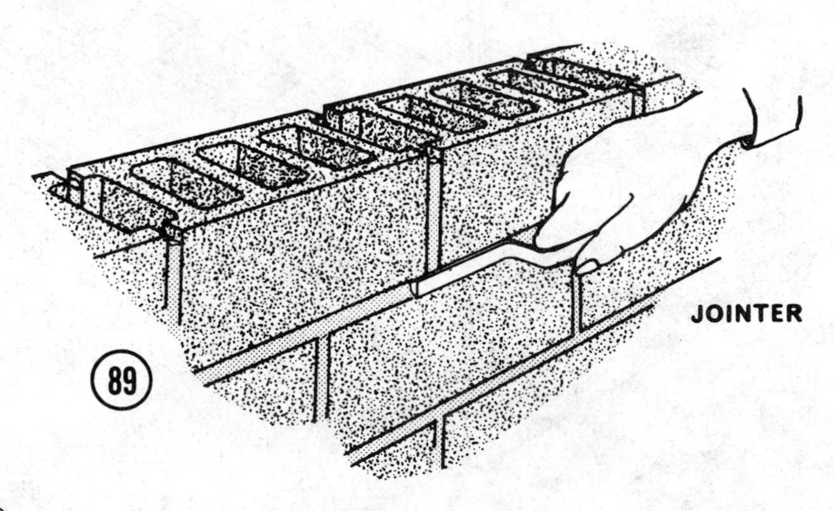

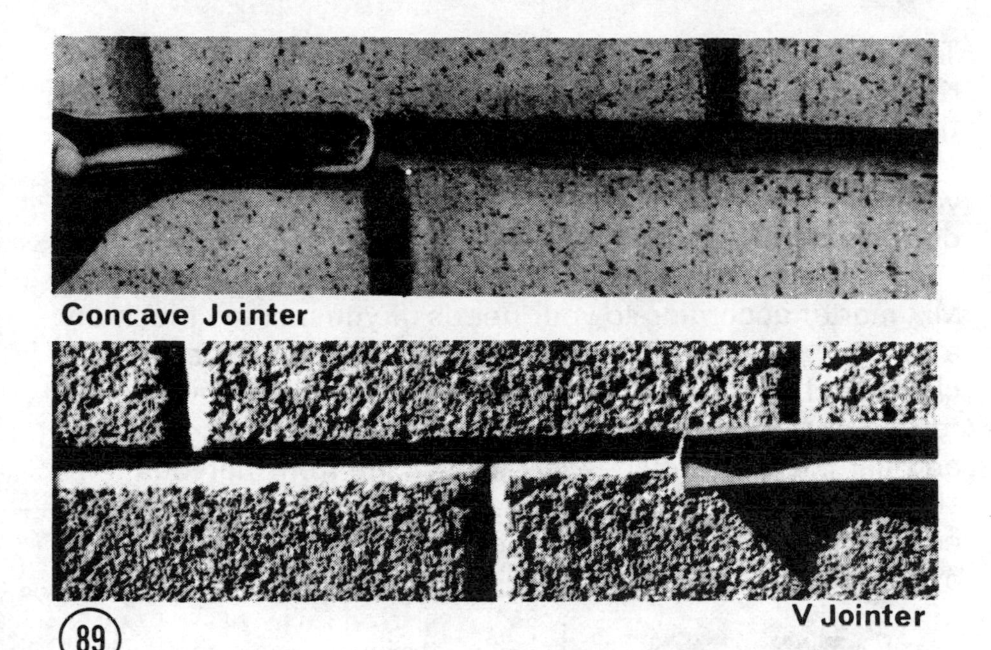
Concave Jointer

V Jointer

89

Many masons build up corners to guide line height, Illus. 90, then stretch a line and fill in each course. It's always necessary to check each block with a level horizontally as well as vertically.

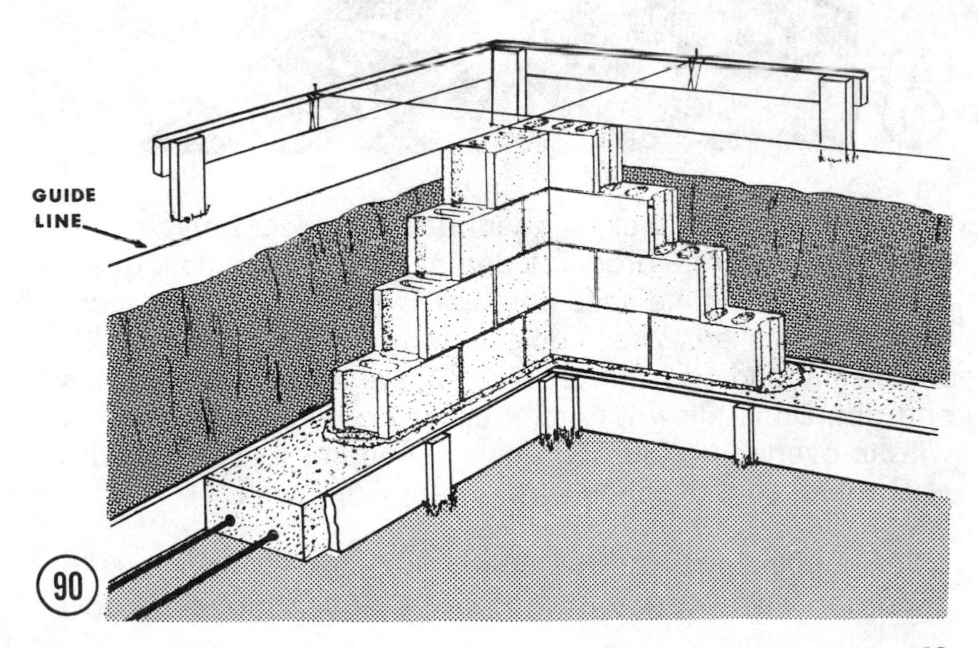

90

The key to laying block like a pro rests in the use of the level. Keep it clean at all times. Don't misuse it. Wash off any particles of mortar. Keep it and your trowel store fresh and slightly oiled. Keep a pail or barrel of clean water handy so you can wash all tools the moment you stop working, even during a coffee break.

Mix mortar according to your needs. If you mix too much on a hot day, it requires a lot of remixing. It also creates a psychological need to rush. And this is something you should refrain from doing. Laying block requires time, concentration and care. It also provides instant escape from nervous tension. Establish a leisurely pace, remembering you need to condition muscles to lifting block, and your back to bending over.

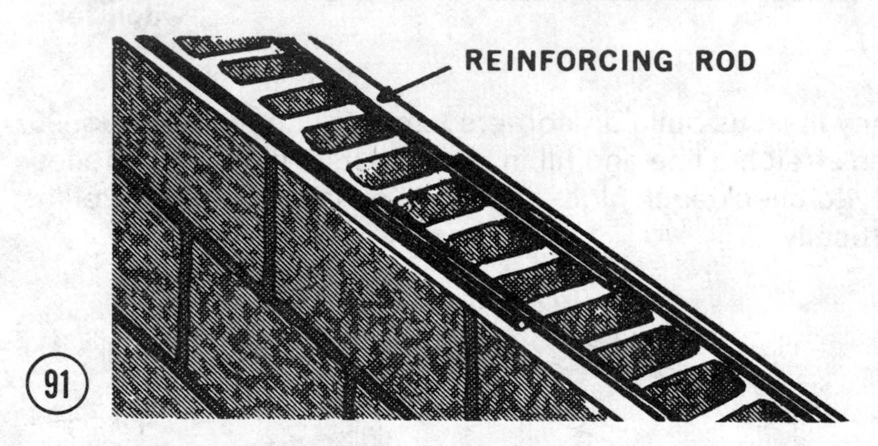

Always mix mortar to exact quantities specified. Don't improvise. Your concrete products dealer can answer most questions concerning the need and use of reinforcing wire, contraction joints, etc. Wipe cup grease on rod or wire, Illus. 91, before embedding in mortar. Keep the wire clean, free of dust and dirt. Don't allow it to pick up any foreign matter. Allow wire to overlap 6″, Illus. 92. Always bend at least 6″ of wire at corners, so it overlaps wire on adjoining wall.

When building an L-shaped foundation, Illus. 93, always start at corner A and work toward B and C. If in doubt as to how blocks will lay out, make a dry run. Position blocks ⅜″ apart. When you lay them up you can then widen a joint or two or use a half block where needed.

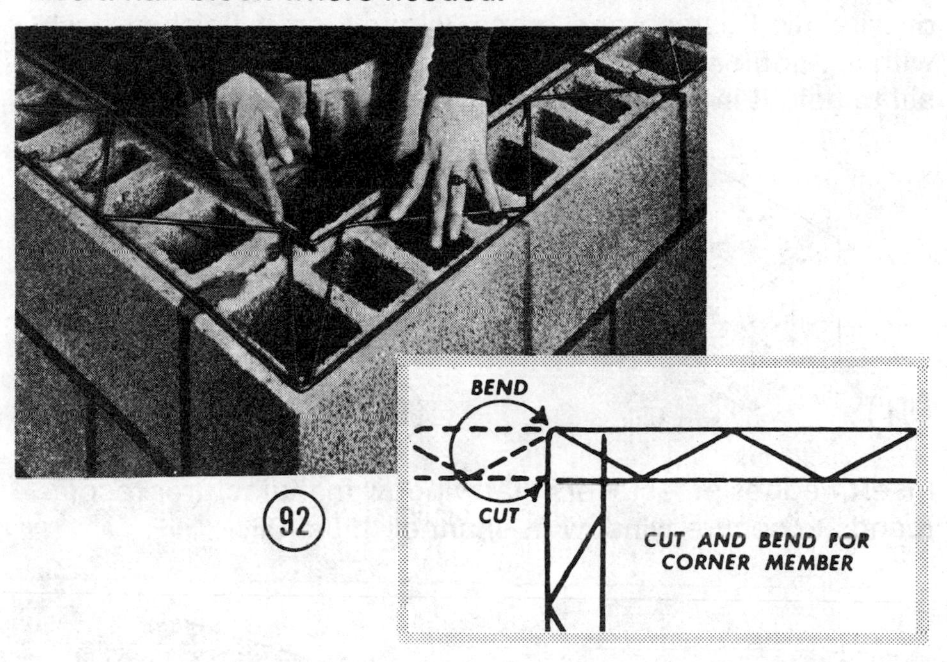

92

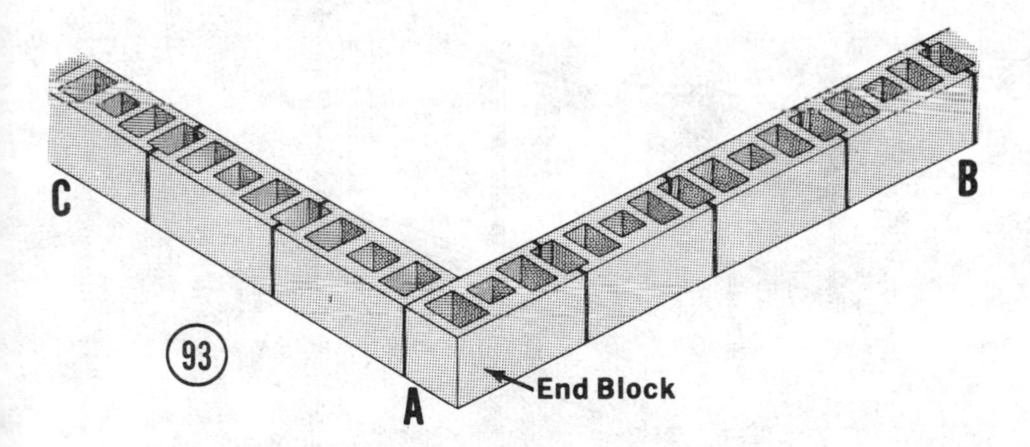

93

Always center joints over previous course, Illus. 86. Never allow joints to line up, except where a control joint, Illus. 108, is required. Under certain conditions control joints are required on a long or high wall to relieve contraction. This is especially important in walls over 30 ft. in length.

Always use channel blocks, Illus. 94, at window openings for steel or aluminum windows. Use half blocks, Illus. 95, where required. Lay up blocks on one side of opening to full height of window. Position and brace window so it finishes flush with top course of block. Use pieces of block or brick under sill to hold it in position.

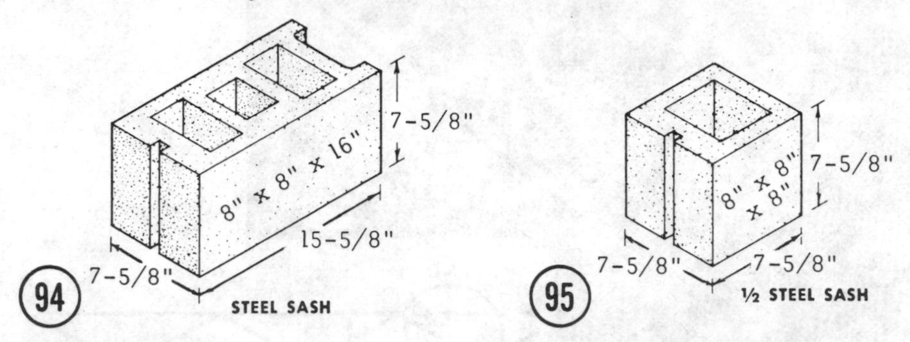

Insert wedges or fasteners that window manufacturer recommends to secure window in channel, Illus. 96.

Build a sill form, Illus. 97, 105, and pour sill, following directions outlined.

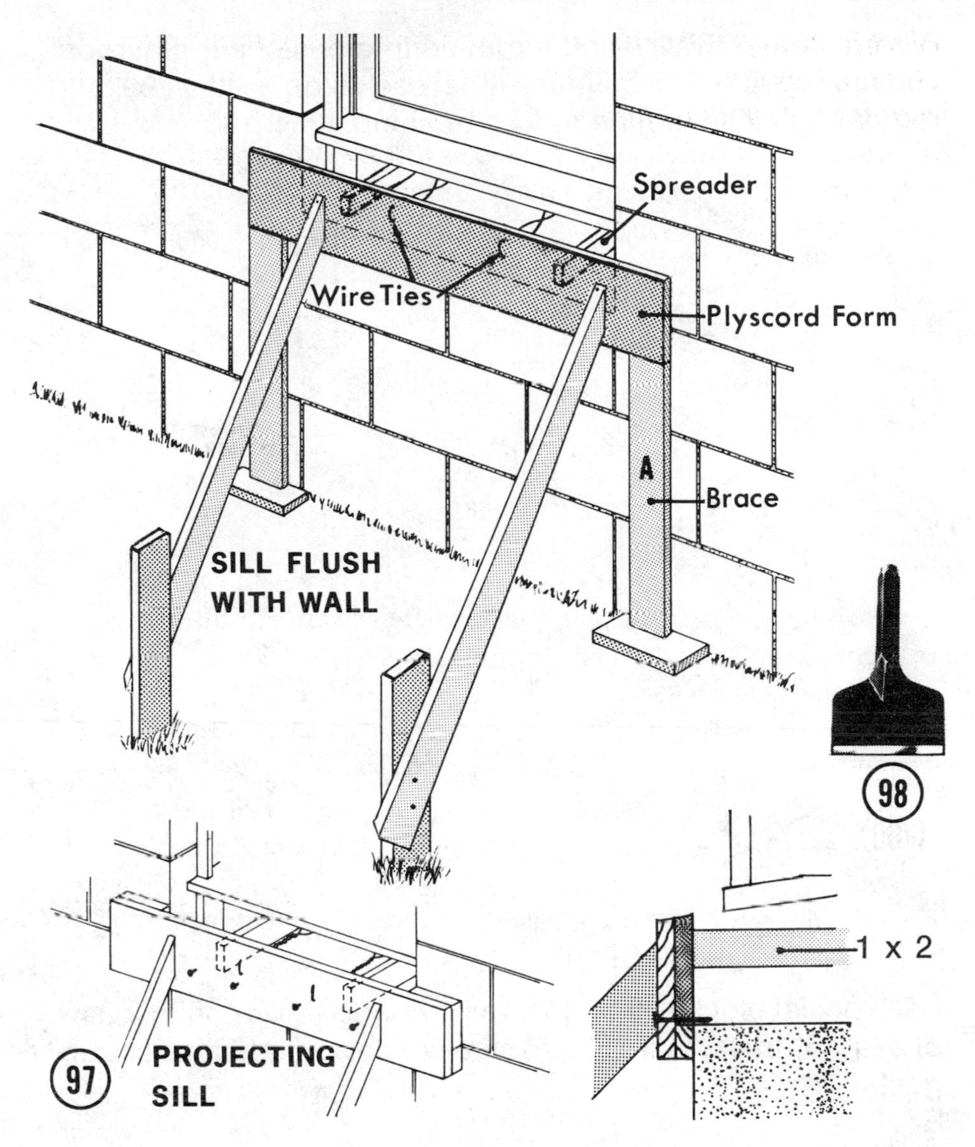

Use finished end blocks, Illus. 75, on all exposed corners. Fill in each course with full or half blocks, or cut a block to size required. Build wall up to beam height.

If you need to cut a block use a chisel, Illus. 98. Draw a line on block. Place block on a level surface or on a flat bag of sand. Strike chisel with a hammer working your way along the drawn line. Be sure to stagger joints, Illus. 86, except where noted on page 79.

Where a block wall is left exposed in a basement playroom you may want to use bullnose steel sash blocks, Illus. 99, and regular bullnose around a door opening, Illus. 100.

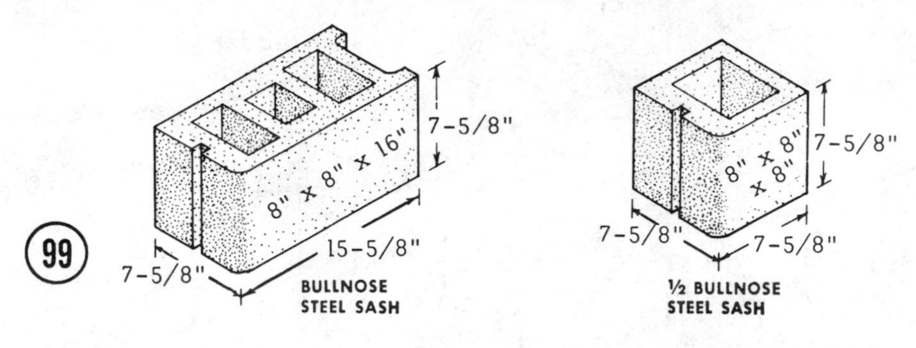

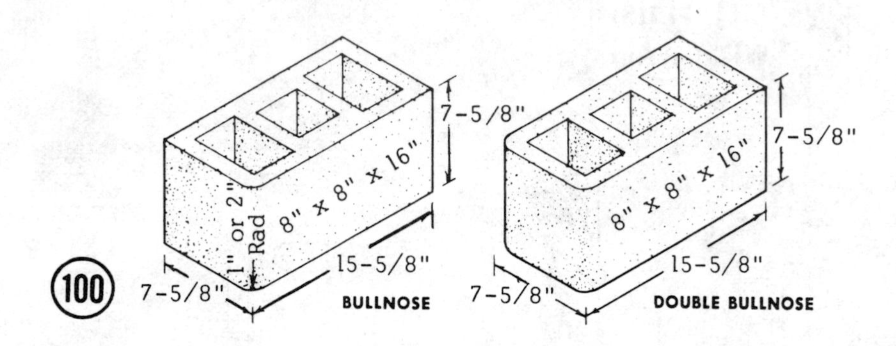

Use special blocks, Illus. 101, for a wood sash, or build a 2 x 4 or 5/4 x 6″ buck, Illus. 102, to size window manufacturer suggests.

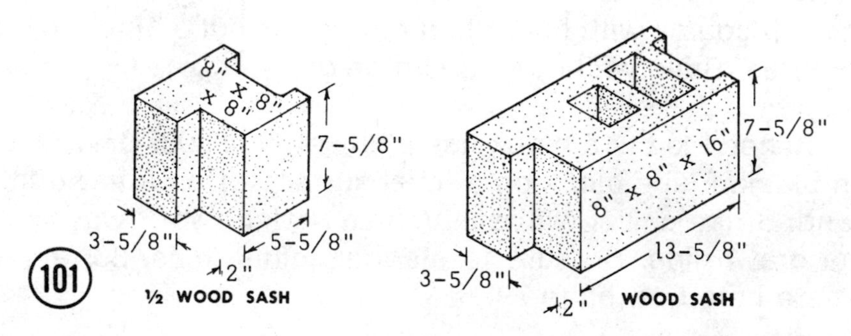

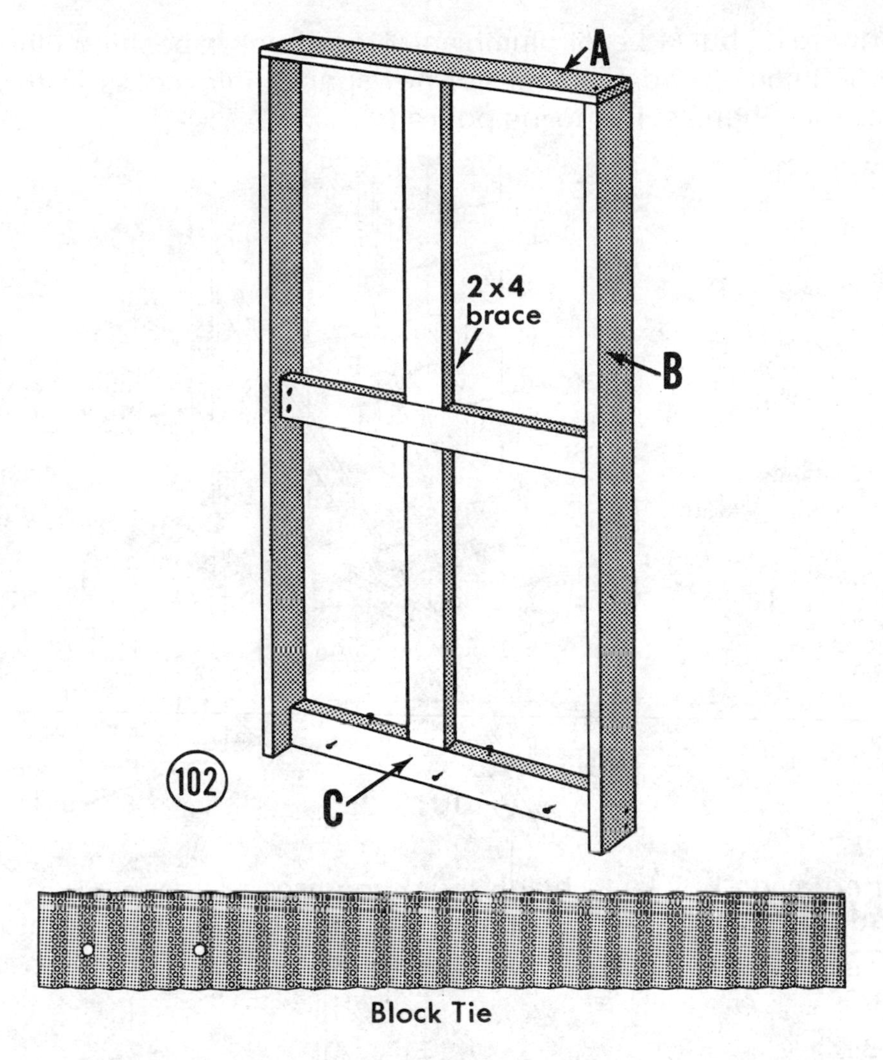

If you can't purchase a precast concrete sill, follow procedure described below.

A buck built to size window manufacturer suggests, positions window in opening so buck finishes flush at top with a course of blocks, Illus. 103. It also simplifies "pouring a sill" after setting window. Nail A to B. Nail B to C. Use 2 x size lumber window requires for C. Drive some 1″ big head nails into side and top of C, Illus. 102. Nail block ties to B at height needed so each can be bent to lay into a mortar course in block, Illus. 103.

Square up buck. Level, plumb and brace buck in position, Illus. 103. Ruffle up and push balls of newspaper into cores in top course where sill is to be poured.

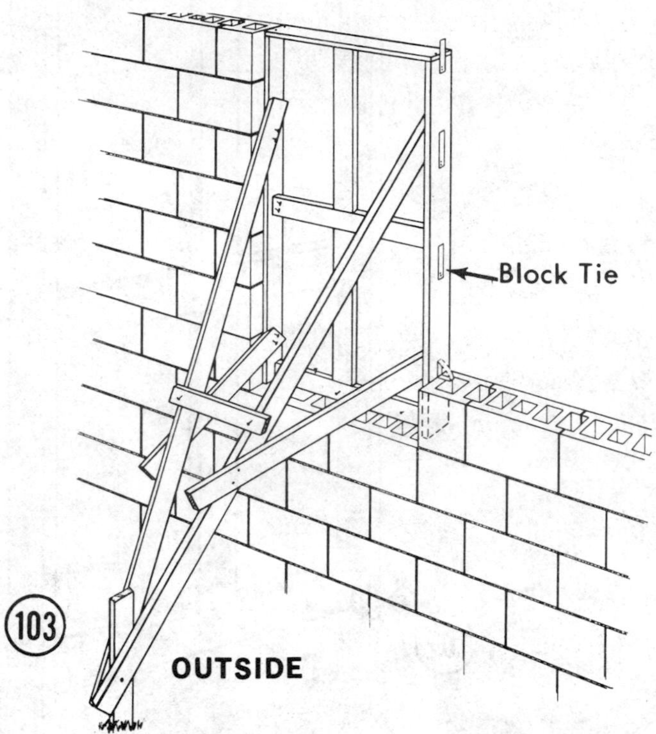

After laying blocks to height buck requires, remove braces and place window in buck, Illus. 104. Nail through casing into buck. Test open window.

Use finished end blocks to build wall up to top of window. Allow ¼″ or spacing that window manufacturer suggests between block and buck. Be sure to bend ties down into each or every other mortar joint as you lay each course of block. Fill joint between block and buck or between block and window frame with non-hardening calking, Illus. 96.

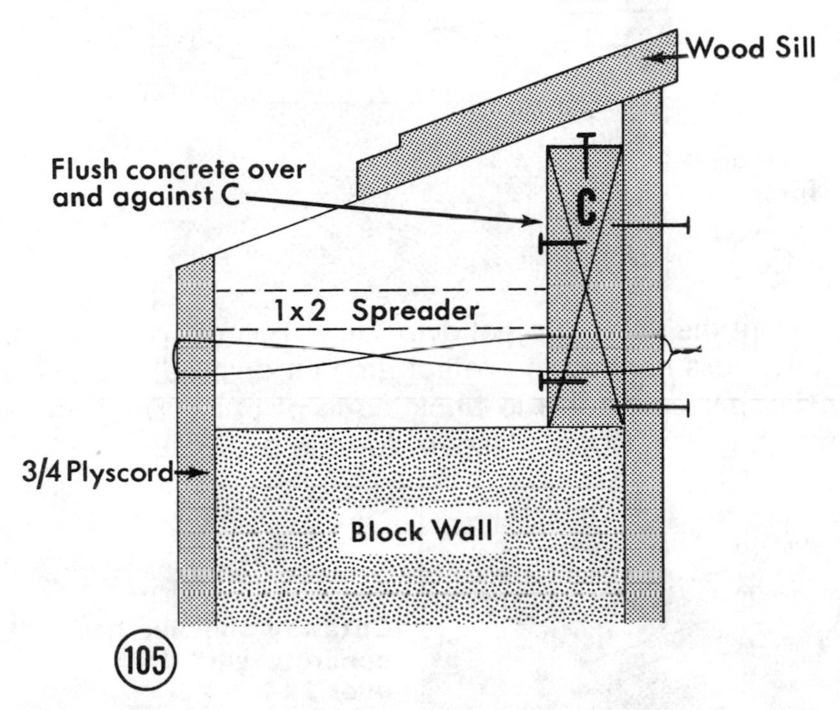

To build a sill that finishes flush with face of blocks, Illus. 97, 105, use ¾″ plyscord for a form inside and outside. Plane top edge to angle required. Support form in position with 2 x 4 legs A, and angle braces to stakes. Nail form temporarily to inside edge of buck. Don't drive nails all the way. Drill holes and use a twisted wire tie and 1 x 2 spreaders to hold sill form together and apart, Illus. 105. Mix one part Portland cement to three parts sand for a mortar mix. Use color cement that matches blocks. Remove spreader as you fill form. Work mortar up under window sill and pack it in. Trowel finish sill to pitch shown. Allow sill to set at least four days. Cut wire and remove form. Recut wire, plaster over end.

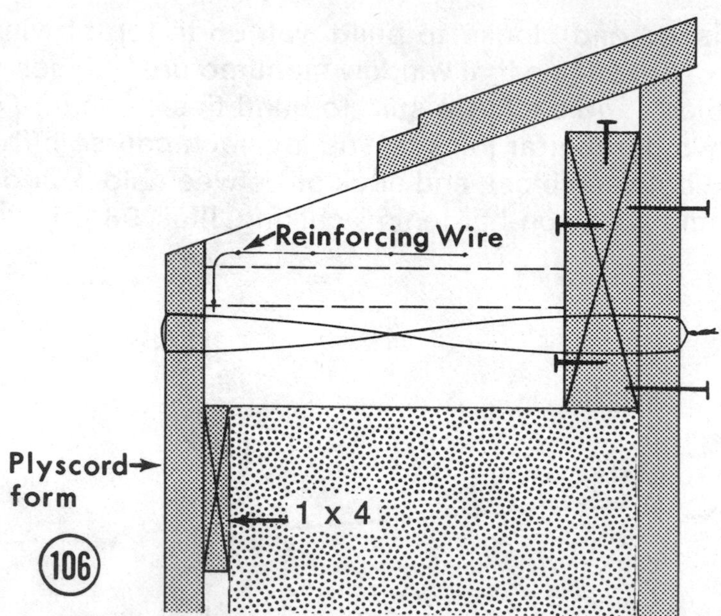

If you want the sill to project over face of block, place a 1 x 4 or thickness needed to project amount desired, Illus. 106, 107. Nail plyscord form to back edge of buck. Don't drive nails all the way. Brace plyscord form in position.

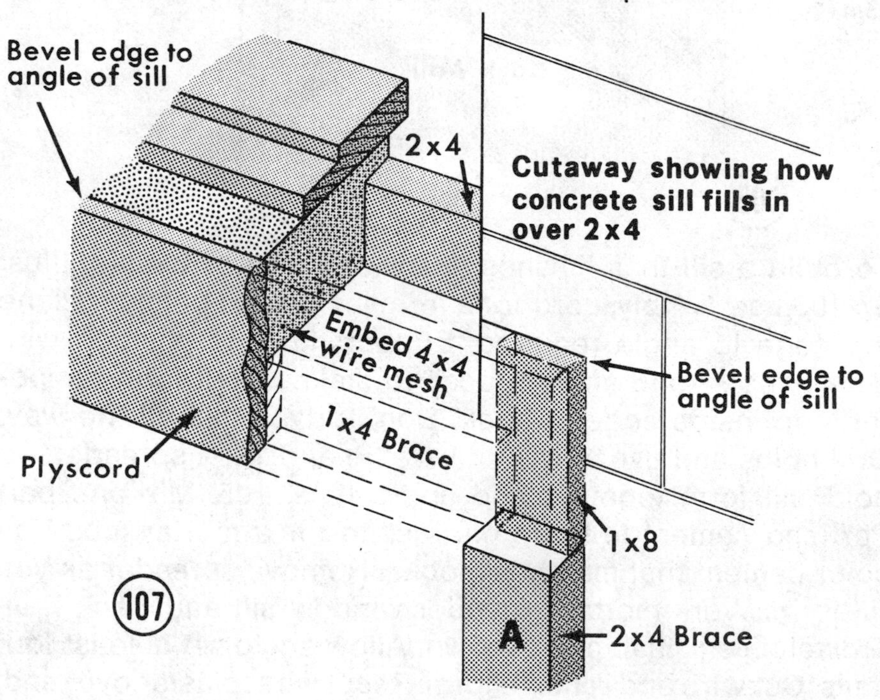

CONTROL JOINTS

Due to the many different conditions, i.e., size of block, height and length of wall, pressure from a slope, etc., etc., that can arise when building a block wall, or pouring a foundation, the information concerning control joints is offered only to alert the reader to specific points of construction. This encourages a more intelligent and understanding discussion when you explain your job to a readymix or concrete products dealer. Non-continuous reinforcement, in every other course, permits lateral transfer of movement. Grease the wire before embedding. The wire helps distribute stresses that could crack a wall. Where a build-up of pressure might produce a large single crack, the wire helps absorb and distributes the pressure, and, in severe cases, produces only small cracks. Bend at least 6″ of wire at ends and lap it over. The Bureau of Standards recommends the use of horizontal reinforcing wire in every other bed joint in walls exceeding 12′ in length. Also in the mortar bed of blocks under sills, Illus. 104. Use #6 or #8 cold-drawn wire.

One method of relieving pressure is through the use of steel sash jamb blocks, Illus. 94, and a preformed rubber control joint, Illus. 108. Install control joints only where readymix or concrete products retailer recommends same.

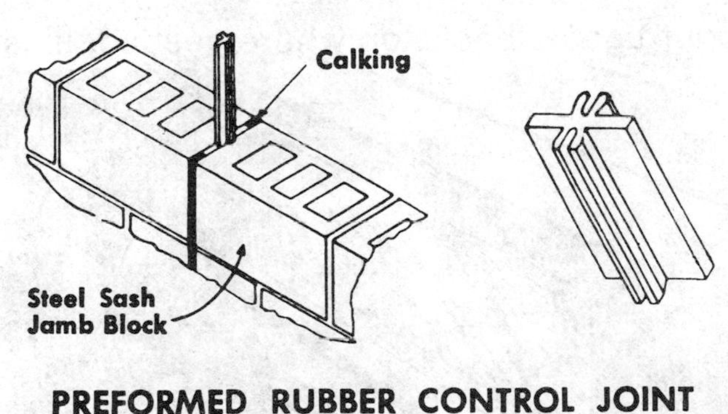

(108) PREFORMED RUBBER CONTROL JOINT

Build footings to height that permits using one size of block. Where necessary, use a course of 4″ block.

If a block foundation is being built at the base of a steep slope, or you are building a high foundation wall, fill cores with concrete as you lay each course. Here again discuss your job with the concrete block dealer and follow his advice.

If your foundation requires a girder, Illus. 109, position blocks in top course and build piers to height girder requires.

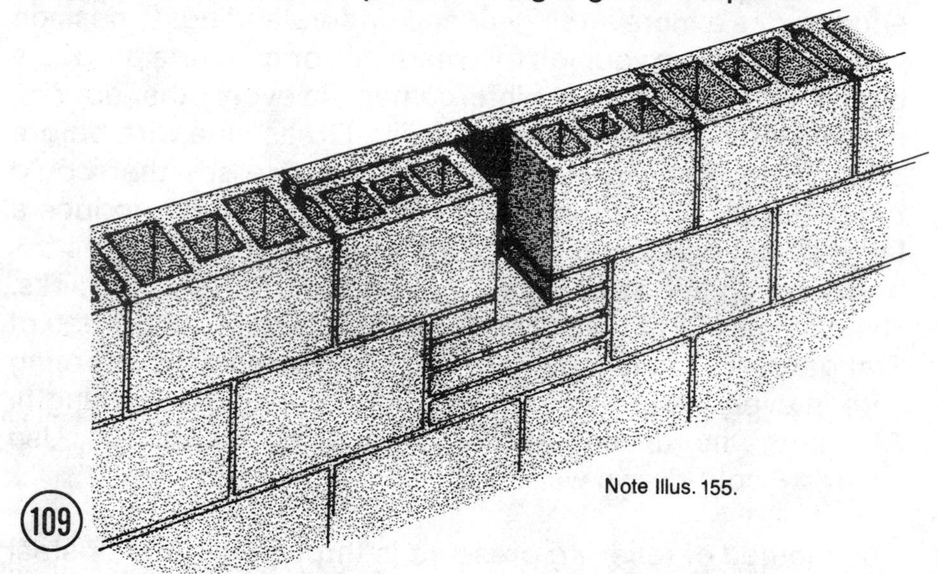

Use channel blocks, Illus. 94, on sides of opening for a basement window. Slide window into channel or follow manufacturer's directions.

Install a courrugated backstop where needed, Illus. 110.

Due to the wide variance in design and size of steel, aluminum and wood windows currently being manufactured, the step-by-step directions offered herewith only provide a guide. It's important to follow directions the window manufacturer specifies concerning opening size, framing, nailing or use of anchoring fasteners supplied. Always allow exact amount of space manufacturer or dealer suggests for calking. Install flashing exactly as manufacturer suggests.

After windows have been blocked in, you can either build a form and pour a reinforced perimeter beam; purchase a precast beam and ask the concrete products dealer to hoist it in position when he delivers same; or use lintel blocks, Illus. 111. Lintel blocks are available in widths that match block sizes.

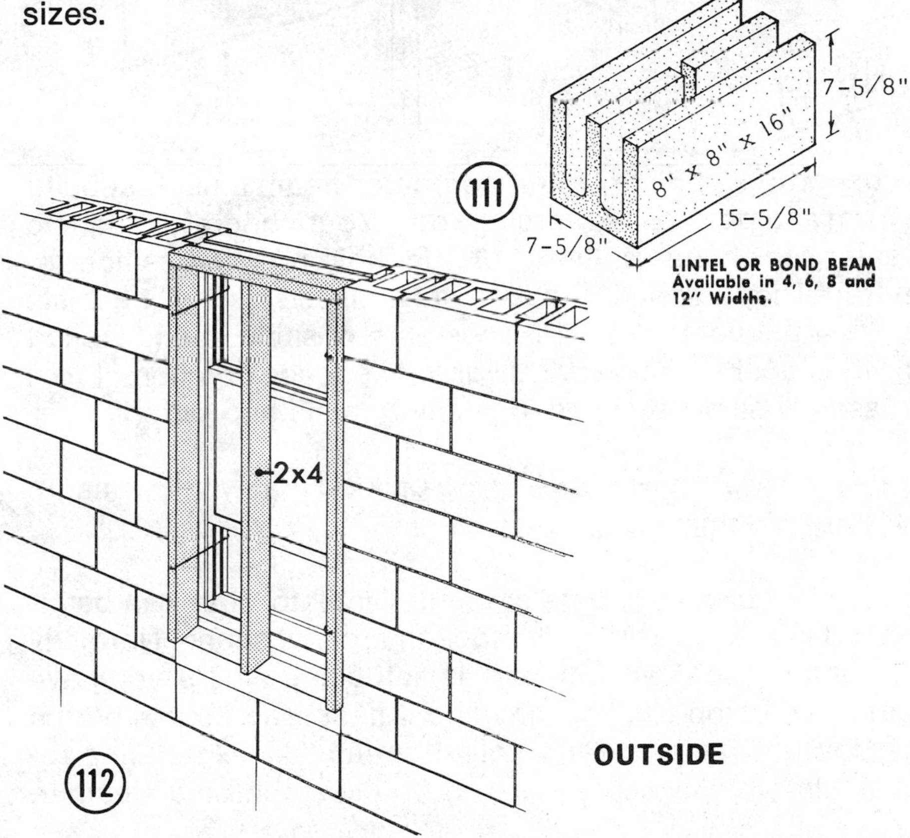

To build a lintel in place, fill recess in front of window with 2 x 4 shoring, Illus. 112.

Place lintel blocks, groove side up, end to end, Illus. 113. Fill grooves with 1″ of concrete and embed ½″ rods full length of assembled blocks. Fill groove with another 2″ of concrete and embed a second rod. Fill block to top. Allow beam to set undisturbed for four or more days. Complete perimeter course with blocks.

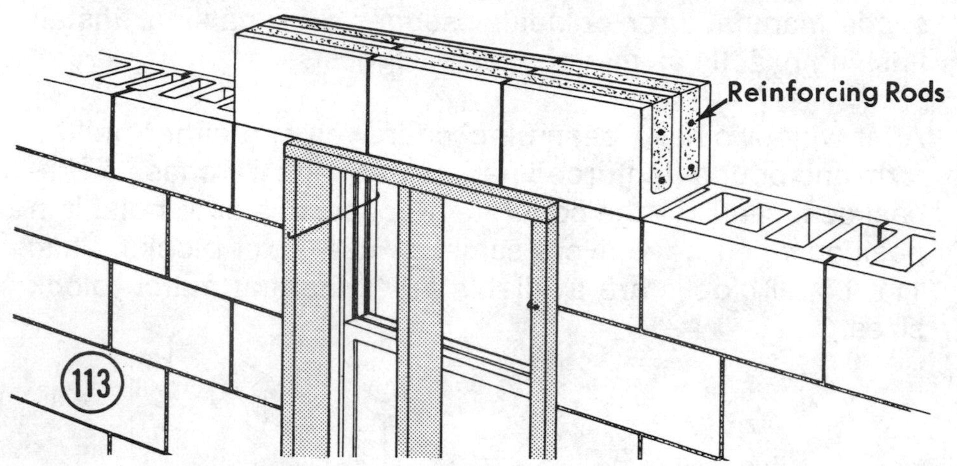

Those who decide to pour a perimeter beam must first build a form and position reinforcing rods. Your concrete products dealer sells hangers, Illus. 114, that hold rods in position required. They also sell 18-gauge galvanized steel ceiling joists or truss anchors, Illus. 115. These are positioned and locked in when beam is poured. This greatly speeds erection of roof trusses, if same are used, or ceiling joists and rafters.

Calk joint around windows, using calking that window manufacturer recommends.

Illus. 116, shows one type of metal ventilator that can be installed in the top course of block in a crawl space. Normally two of these, one at each end, is sufficient. Where you have a moisture problem, you may find it necessary to use both a window and vents in each wall. In areas that experience severe winters, buy vents that can be closed in cold weather.

Wire mesh over a block turned on its side, Illus. 117, can also provide needed air. Use fine wire mesh to discourage bees.

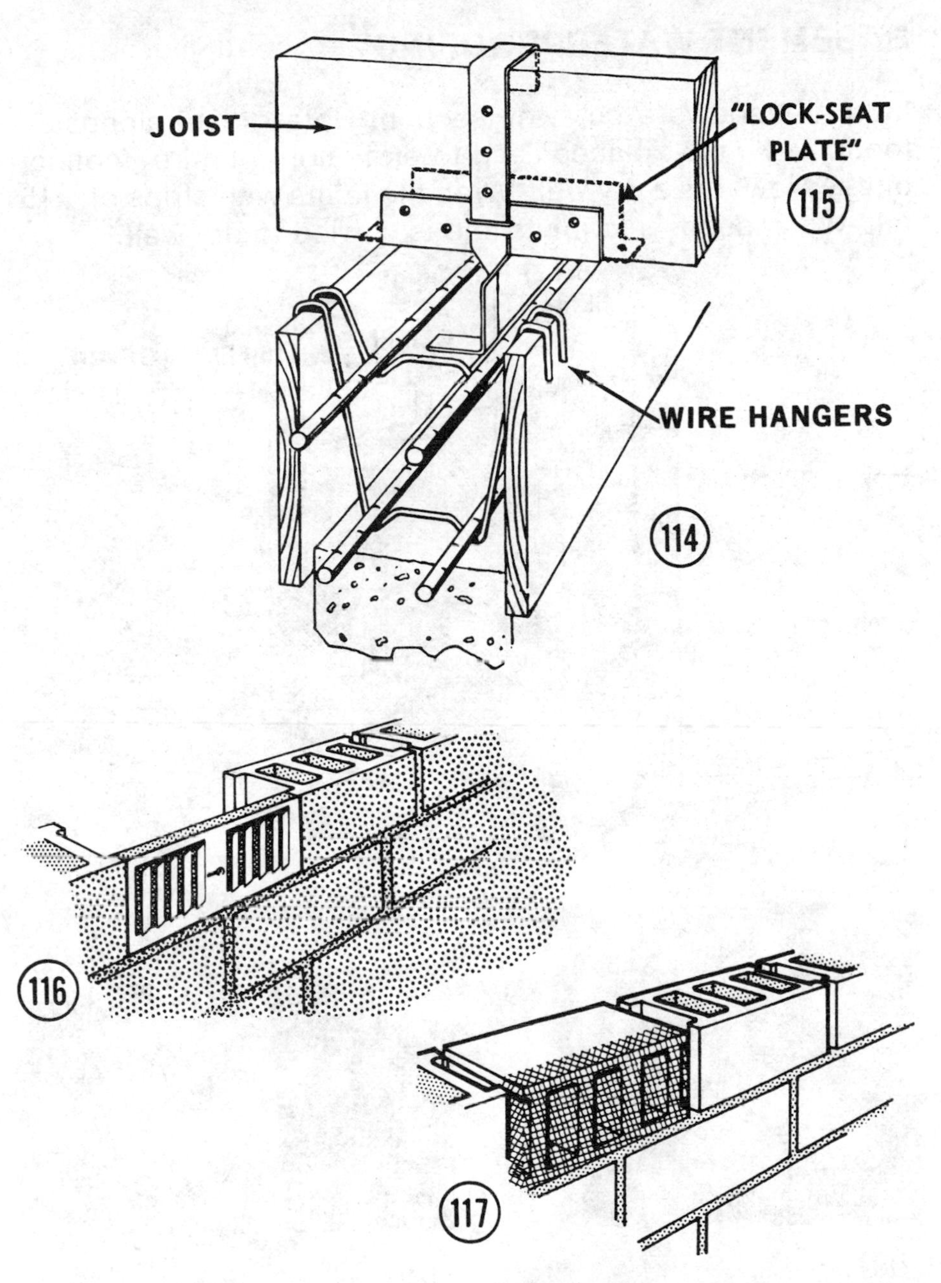

Whether you fill cores in all blocks in each course depends on the pressure exerted against the wall. If there is only normal pressure, only the cores in top course of blocks need to be filled, and these are only filled when you get ready to set anchor bolts or anchor clips.

BASEMENT WATERPROOFING

Illus. 118 shows a cutaway section. Note gravel alongside footing. 4″ tile, placed end-to-end along entire footing, pitches towards a dry well. Wrap tile joints with strips of #15 felt. Water seeps into joints and is carried to dry well.

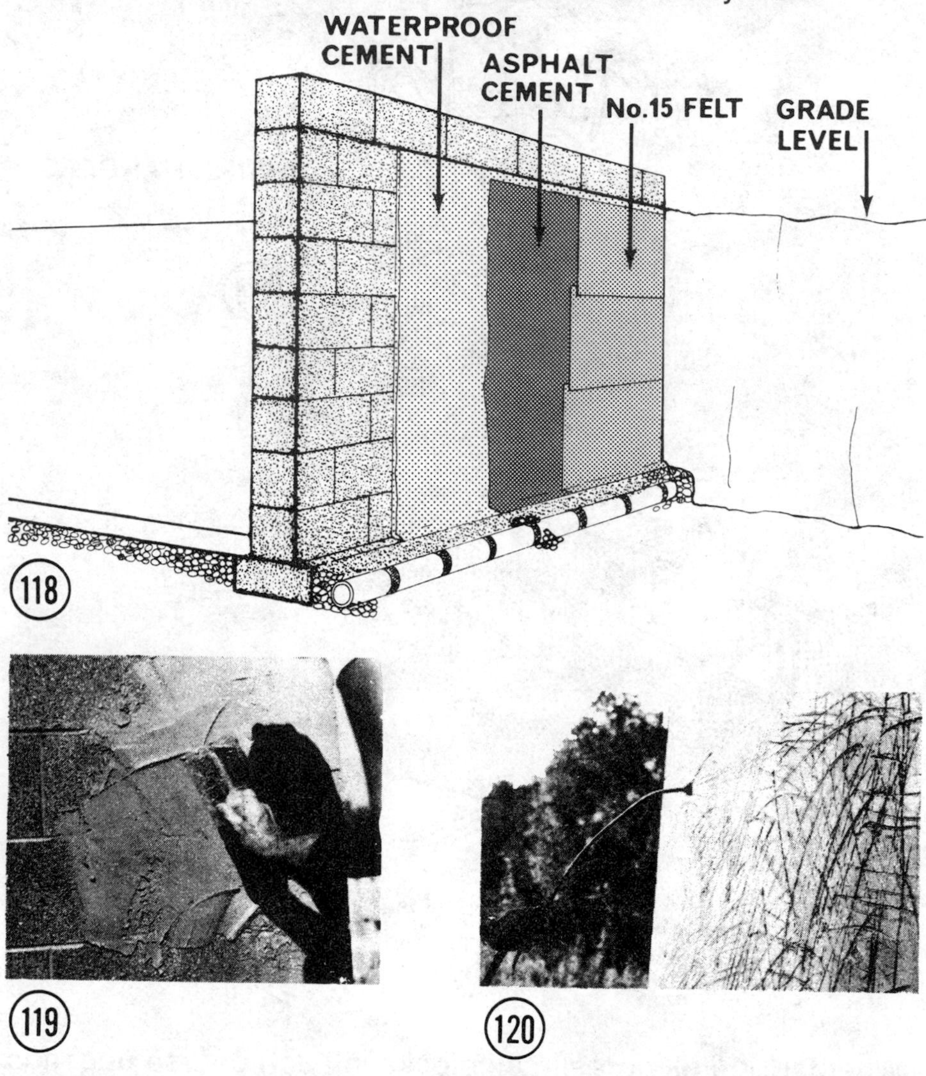

To waterproof a block wall below grade add waterproofing to water prior to mixing one part of cement to two parts of sand. Trowel a ⅜″ to ½″ plaster coat to blocks, Illus. 119. Before it sets, scratch it, Illus. 120. Allow to dry, then apply a second coat, Illus. 121. Spray the surface once a day with a fine mist for three to four days. This helps cure concrete. Allow plaster coat to air-cure for two weeks before applying asphalt roofing or hot tar, Illus. 122. Brush on one coat. Allow to dry, then apply second coat. While wet, embed a layer of #15 felt. Overlap each course 4 to 6″.

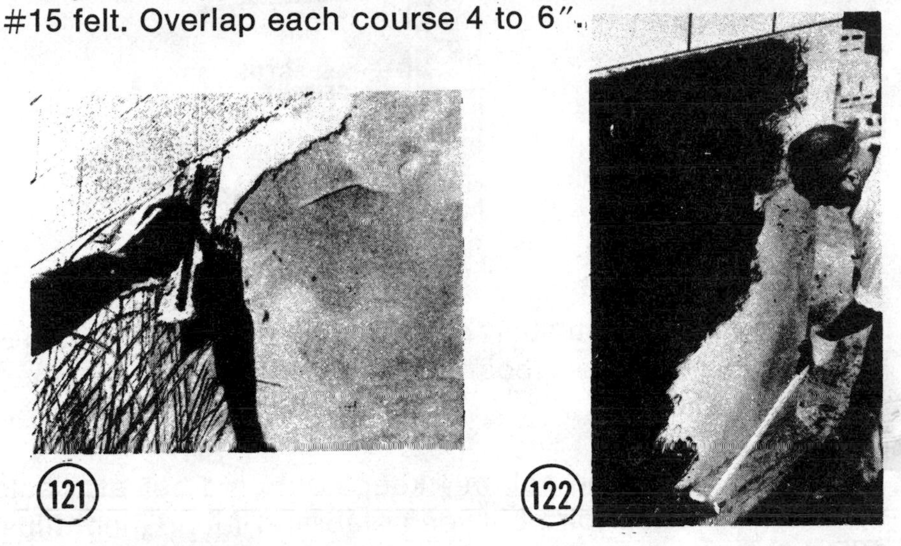

121

122

FOUNDATION INSULATION

Since concrete conducts cold, it's wise to insulate the inside of a crawl space, Illus. 3, 123, or the outside, Illus. 124.

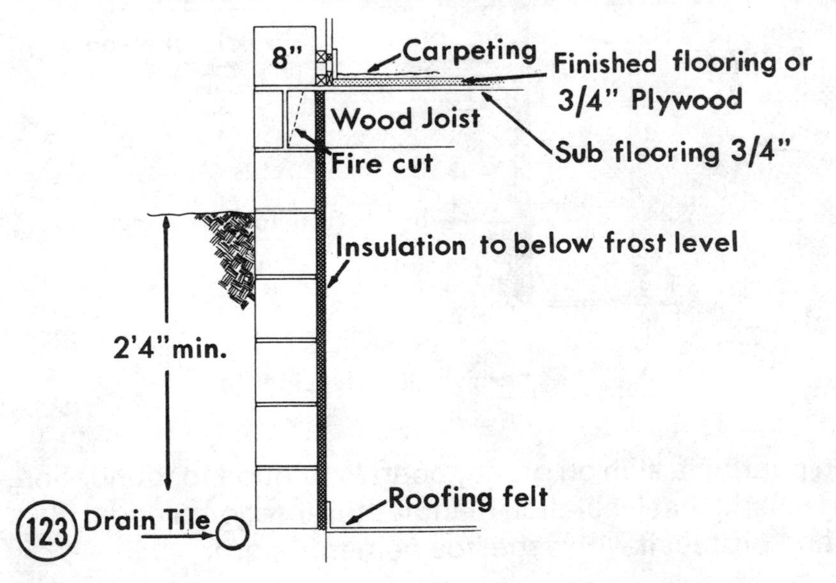

123

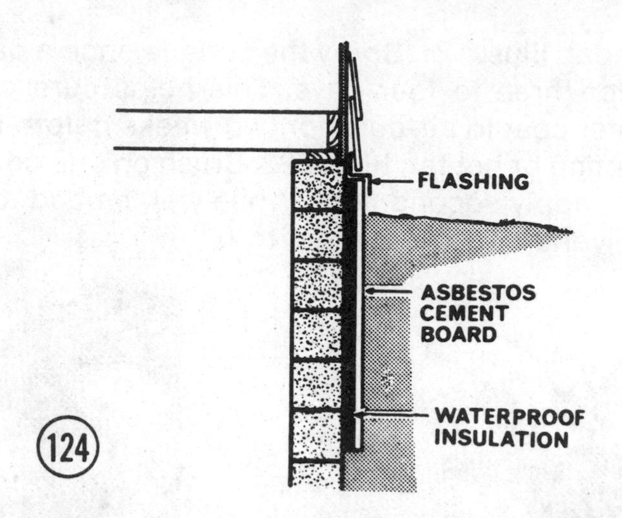

When you apply insulation on outside of foundation, cover with asbestos cement board or other protective weatherproof panelboard.

Rigid foam insulation not only keeps out cold but also acts as a compression joint between slab and foundation, Illus. 125.

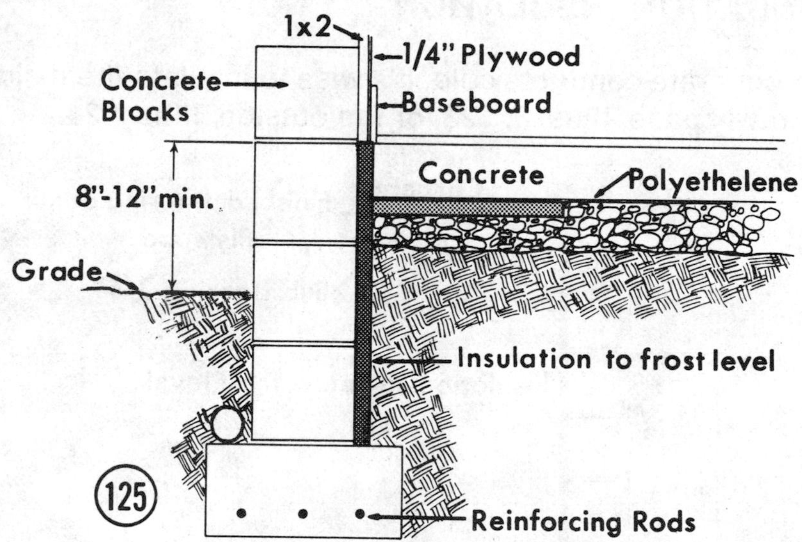

When pouring a slab on grade, bond insulation to foundation, using mastic that insulation manufacturer recommends, Illus. 126, and protect it with asbestos cement board.

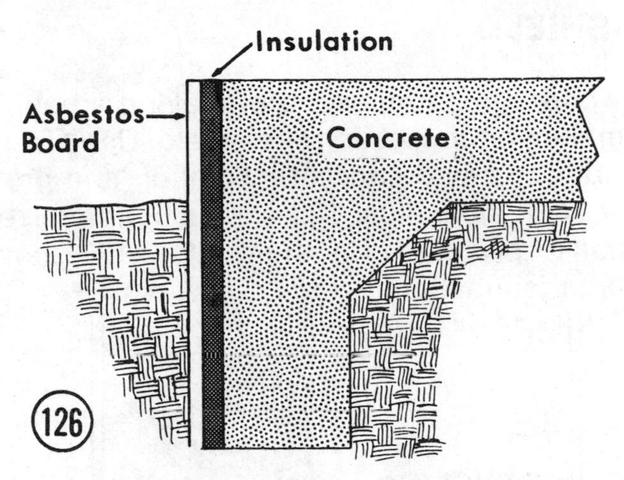

In extremely cold areas, apply insulation to inside of wall to just below frost level, and extend it 16″ to 36″ in position shown, Illus. 127. Cover with gravel. Cover gravel with polyethylene or #15 felt. Forms for the slab are then set up. Wire reinforcing is positioned and the slab is filled in sections.

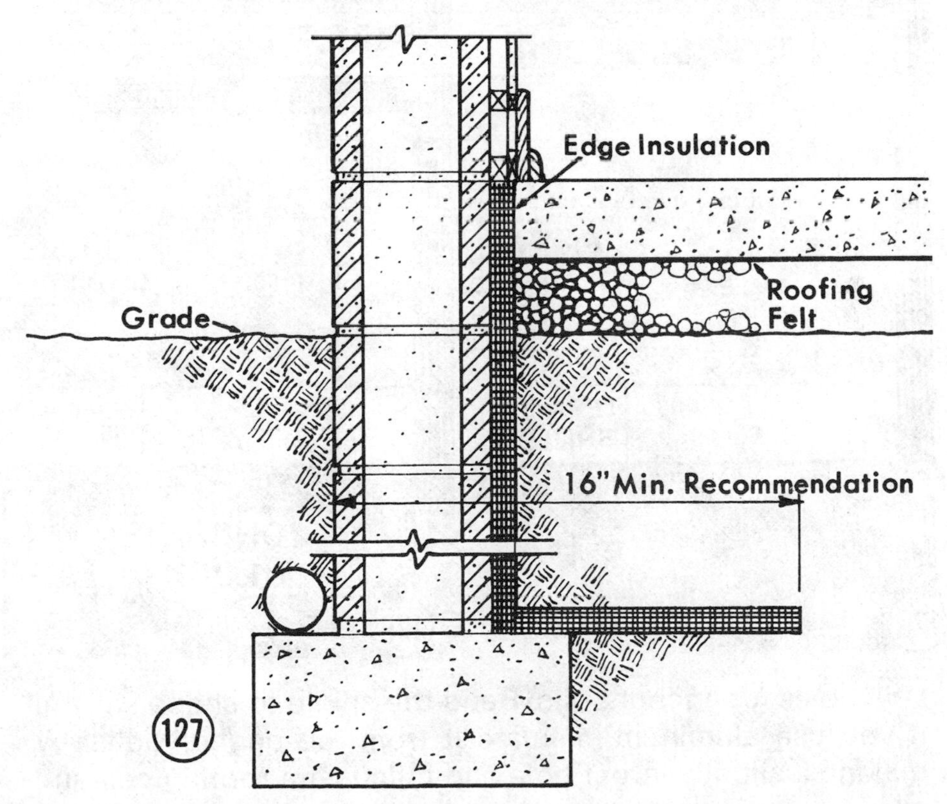

TERMITE SHIELD

To protect wood from termites, before laying a bedplate, install an aluminum or copper termite shield. Use 12″ or width flashing needed to cover edge of insulation and provide an inch lip over each side. Illus. 128 shows bedplate recessed amount equal to plyscord sheathing. Illus. 129 shows bedplate flush with foundation.

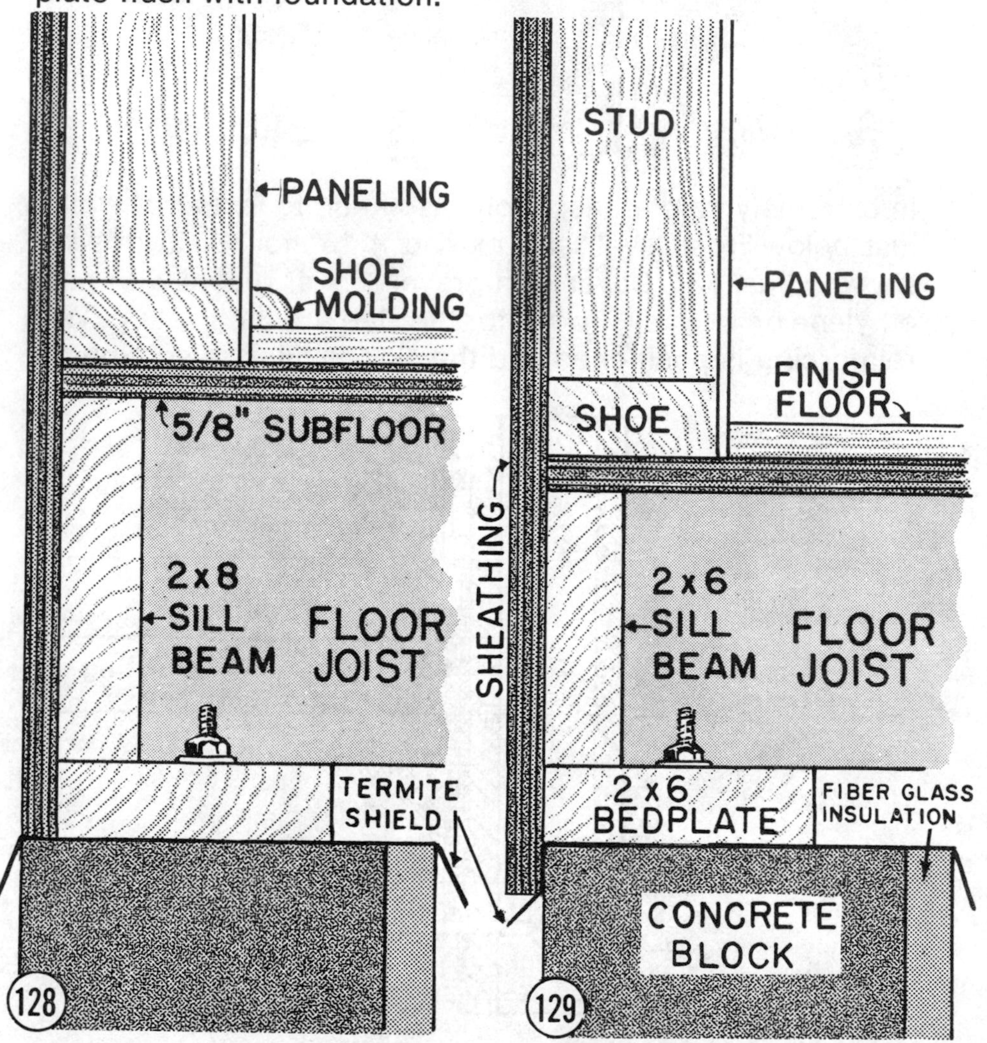

Drill holes for anchor bolts. Bend the shield to shape shown. If you use aluminum, insulate it from the anchor bolts by making a slightly larger hole and filling with roofing cement.

SPECIAL CONSTRUCTION PROBLEMS

To relieve the horizontal stress that results from movement in a large area of concrete, always use a separate footing form under a steel column, etc., Illus. 130, and/or between large areas of concrete. If you use ¾″ plyscord for a form, when same is removed the joint can be filled with a non-hardening compression joint filler.

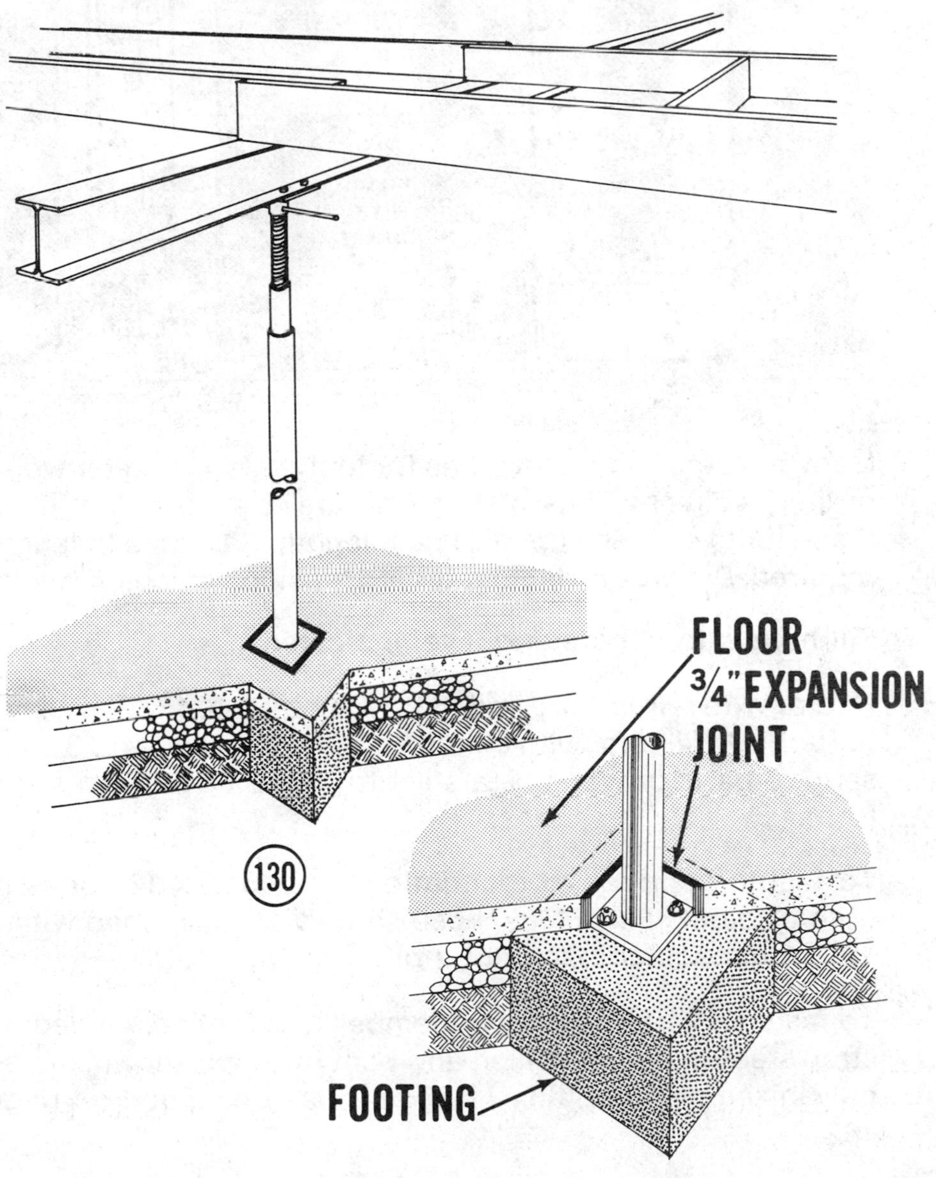

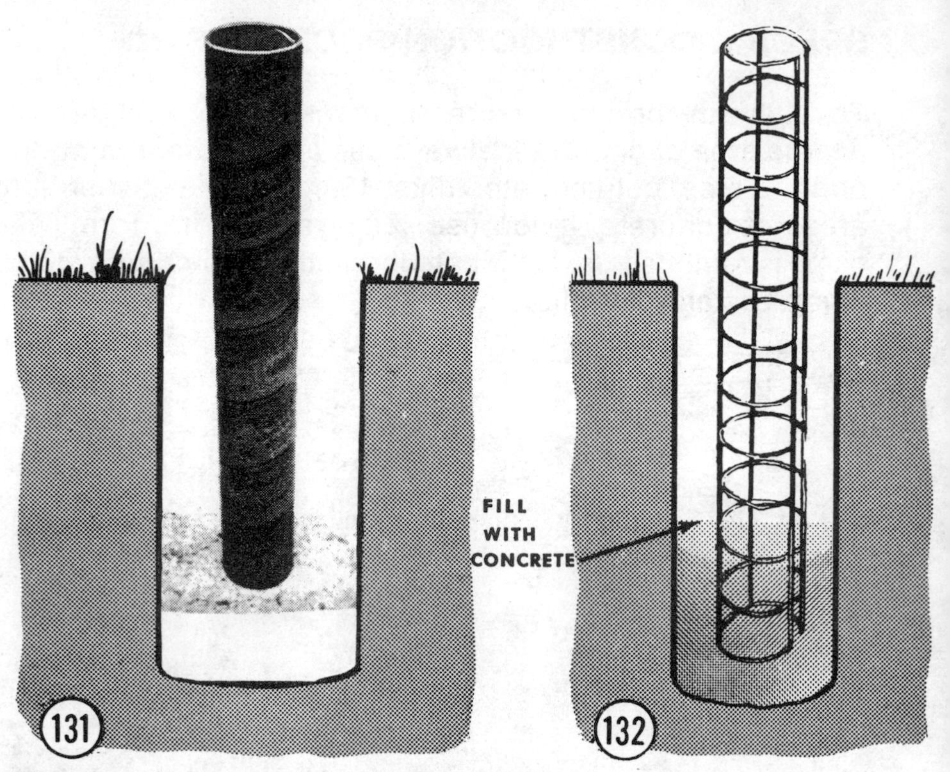

Many builders use round tubes for footing forms under wood or steel columns, Illus. 131. These are available from 6″ in diameter to 4′0″, and up to 20 ft. in length. Saw tube to length required. Dig hole to depth required, position and plumb wire.

Fill hole with concrete to level indicated.

Rod or wire reinforcing, rolled to fit tube, Illus. 132, locks pier to footing. Fill form with concrete mix, 1 part cement, 3 parts sand, 5 parts gravel. Use a stick to eliminate air pockets in form.

To fasten bedplate to foundation, embed ½ x 12" bolts in core of block, Illus. 138. These should be positioned within 12" from each end and not more than 10' apart.

To fasten a vertical framing member to a footing, embed 1″ strap steel, Illus. 133, drilled and bent to shape shown, or use galvanized anchors, Illus. 115, or those described on page 162.

Due to a rapidly shifting weather pattern wherein winds of hurricane force devastate areas previously free from high wind damage, today's construction necessitates using fasteners wherever possible. Since a building is no stronger than its weakest link, anchor piers and foundations to footings, bedplates to foundations, floor joists, studs and rafters following methods outlined. While there is no fully guaranteed method of construction, chances for survival are improved when common sense principles of construction are followed. Always lag screw pressure treated lumber to a foundation, Illus. 133. Note wood fasteners on page 162.

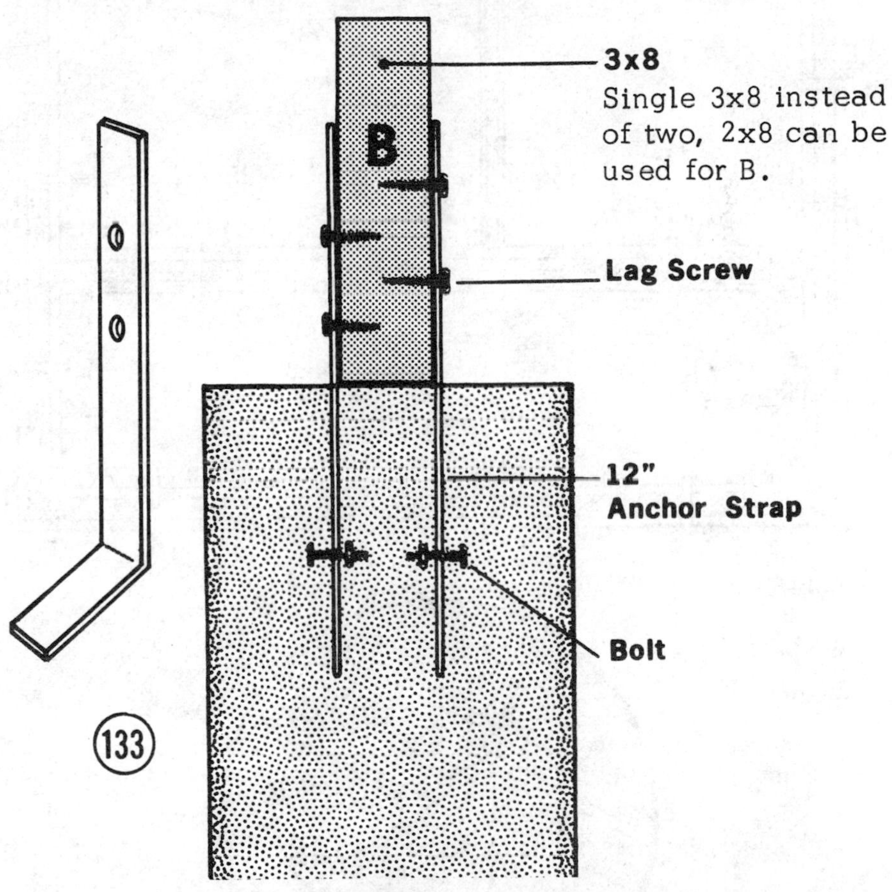

Illus. 134 shows 16″ wide footing forms for a house and porch foundation using 8 x 8 x 16″ blocks, plus footings for a load bearing partition. Always pour a footing for a load bearing partition to support the added weight.

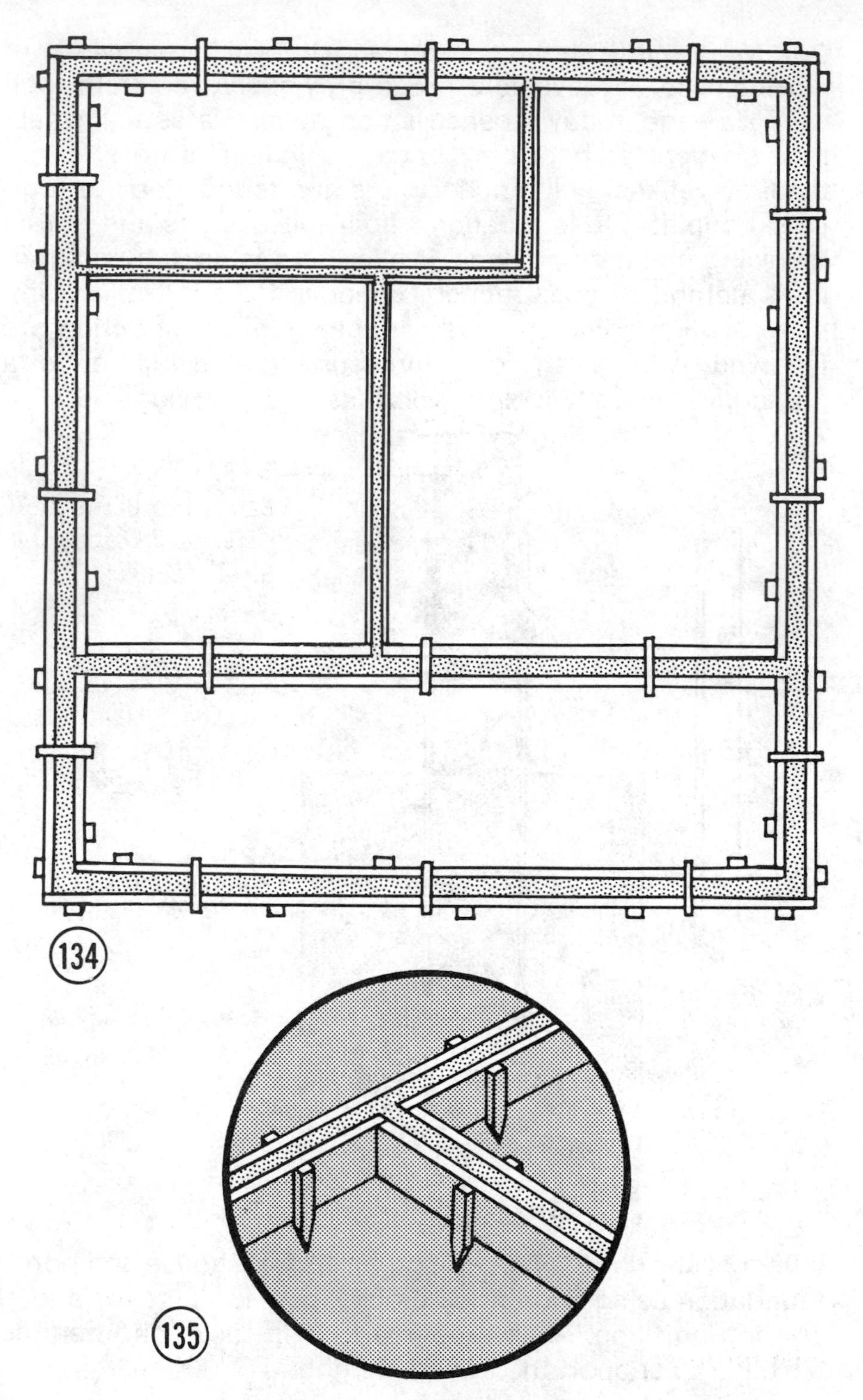
134
135

Build footing forms to size code specifies. 2 x 4's or 2 x 6's, Illus.135, 136, can be used for a basement load bearing partition footing form.

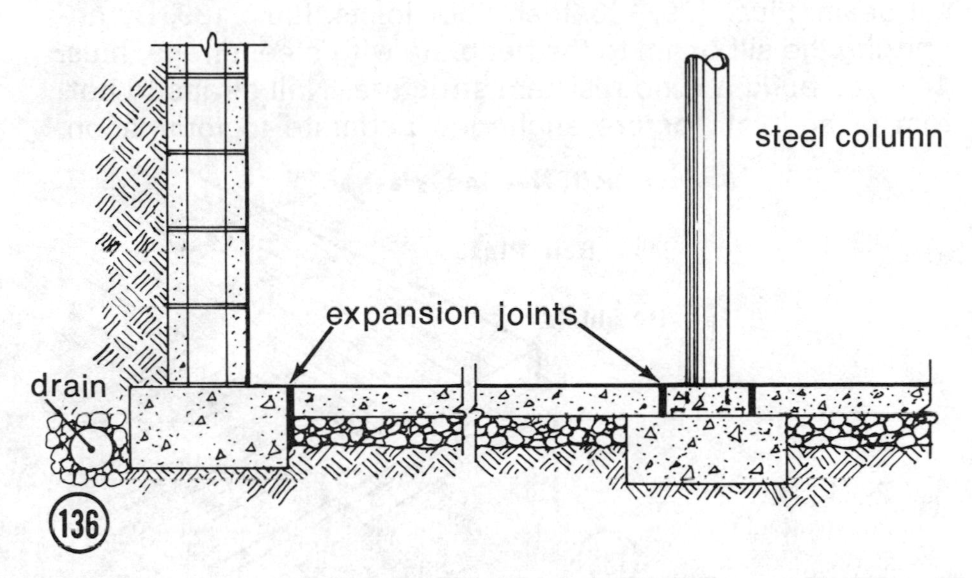

Framing for a basement partition should be secured with "sleeper clips," Illus. 137. This anchors shoe to concrete. Install these at ends of a partition and one every eight feet. The clip is nailed to the shoe with 8 penny nails.

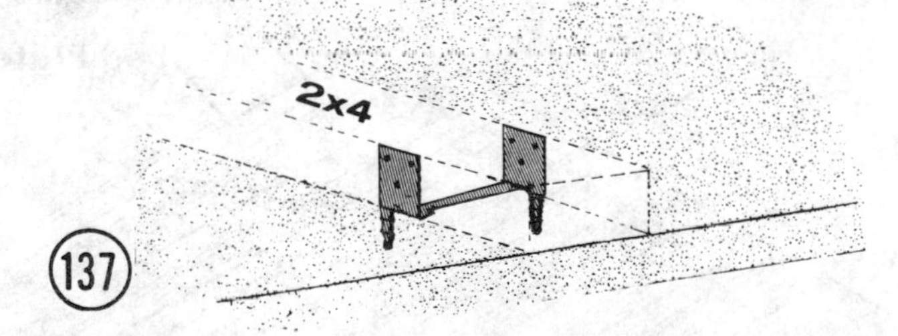

To bolt a steel column to concrete, drill holes in ¾" plywood in exact position holes in plate require, Illus. 28. Nail plyscord to form in exact position. Since a steel base plate seldom requires more than ¾" to 1" of thread above a level footing, screw nut on threads. Be sure to level form for footing and to trowel surface smooth. Always embed a washer on head of bolt to lock bolt in place. When embedding galvanized anchors, many builders insert and bend 8 or 16 penny common nails through holes in anchor, then pour concrete.

FRAMING

Framing starts with the bedplate, a girder where required, sill beam, Illus. 129, 138, then floor joists, Illus. 139. By anchoring the sill beam to the bedplate with steel straps, Illus. 140, you build a wind resistant structure. Nail straps to bottom of bedplate before anchoring bedplate to foundation.

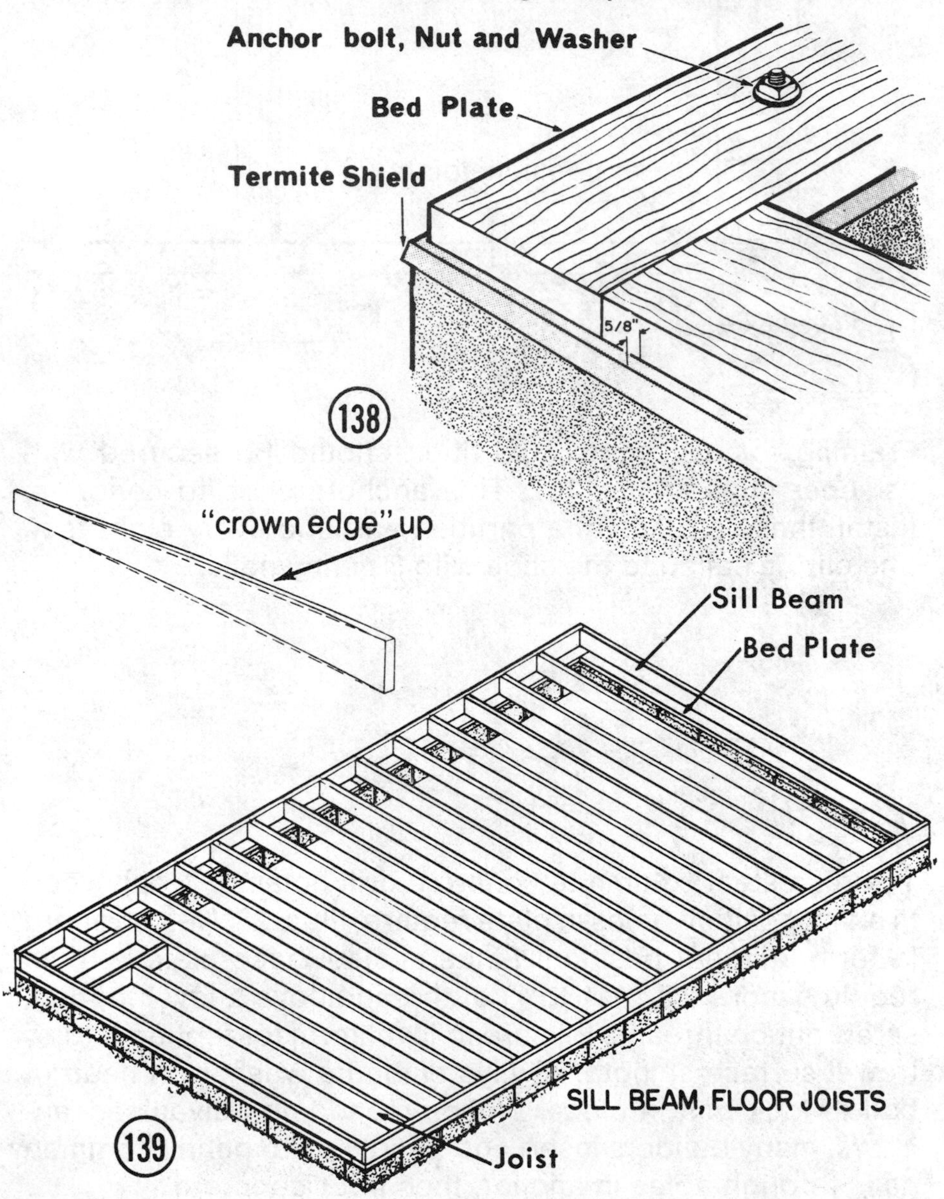

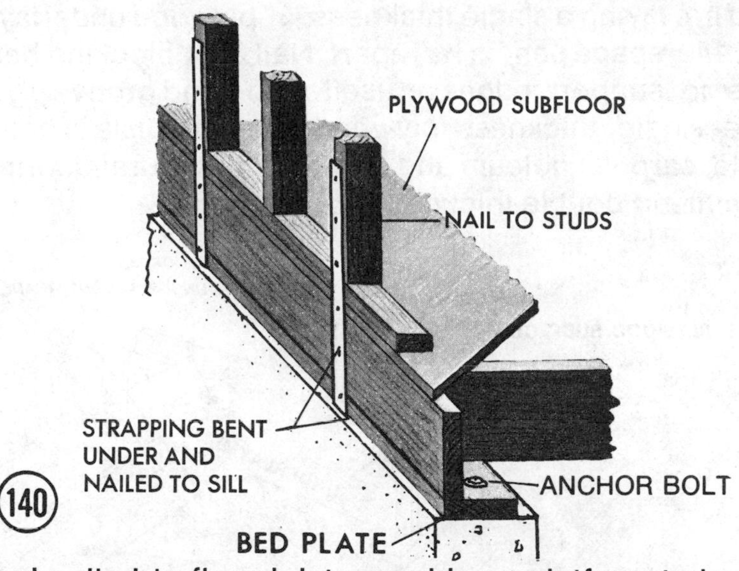

Plywood nailed to floor joists provides a platform to lay out and assemble wall frames. To build a wind resistant structure, apply a ribbon of adhesive to joists, Illus. 141, then spot nail plywood following adhesive manufacturer's recommendations.

If you are laying a single thickness ¾″ plywood underlayment, Illus. 142, space panels 1⁄16″ apart. Nail 2 x 4 blocking between joists to support edges or use tongue and groove plywood. While single thickness plywood glued to joists is approved for tile, carpet, linoleum and other nonstructural flooring, we recommend double thickness flooring, Illus. 143.

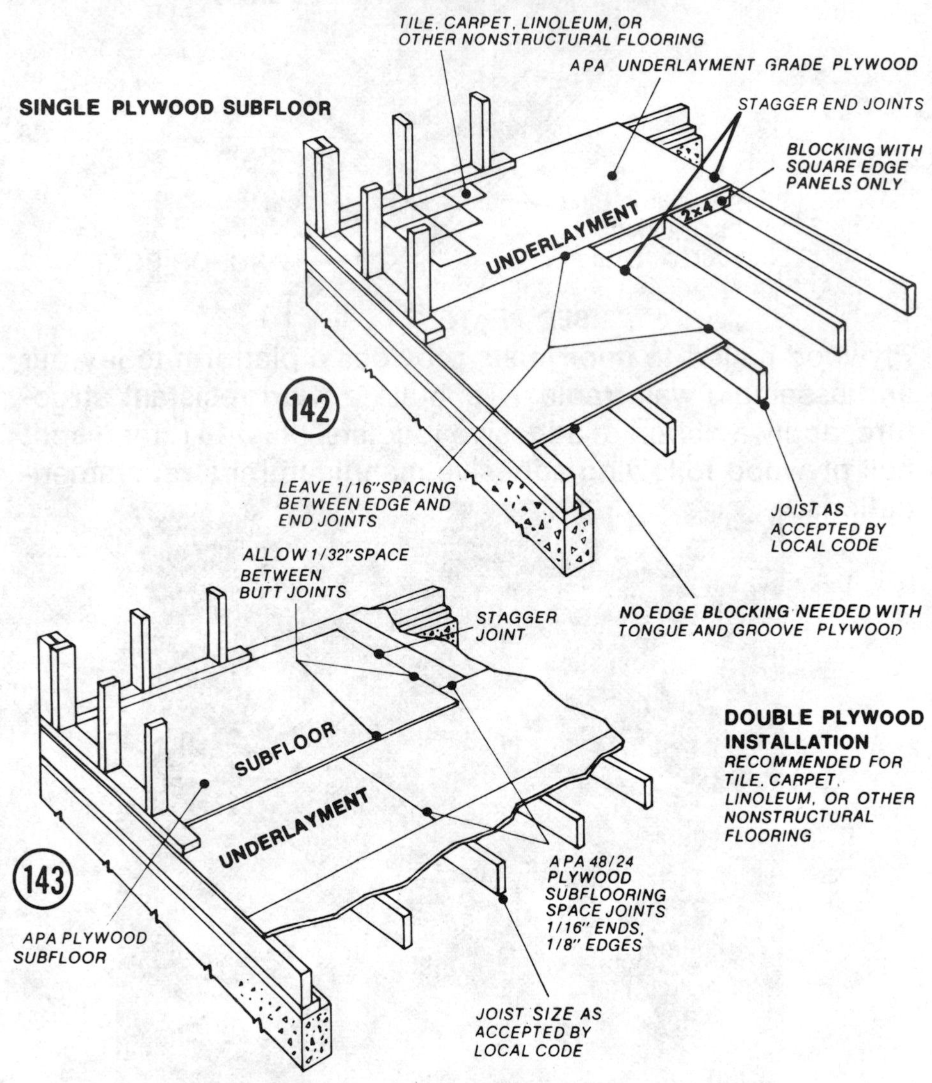

Use exterior grade plywood over plywood underlayment if you want to install ceramic tile, Illus.143. Space top course of plywood 1⁄32″ apart at all joints.

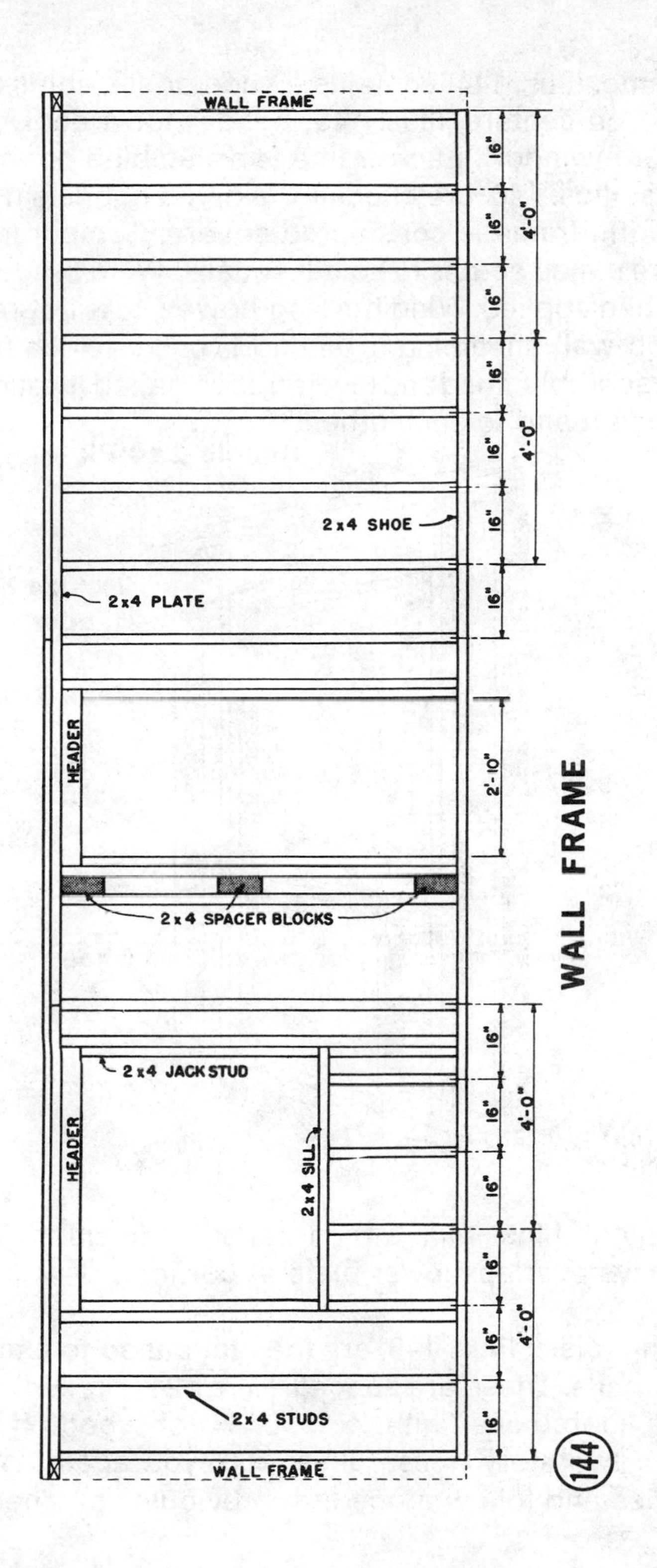
WALL FRAME
16"
4'-0"
2 x 4 SHOE
2 x 4 PLATE
HEADER
2'-10"
2 x 4 SPACER BLOCKS
2 x 4 JACK STUD
2 x 4 SILL
2 x 4 STUDS
144

A wall frame, Illus. 144, contains a shoe or sill, studs spaced 16″ or 24″ on centers, Illus. 145, header for a door, header and sill for a window. Each frame is assembled on the floor. Diagonals, Illus. 146, are checked. When diagonals measure equal length, frame is considered square. Temporary 1 x 4 braces are nailed across to hold it square. Wind braces, Illus. 147, are then applied. Wind bracing, however, is not required if plywood wall sheathing or siding is used. Each frame is raised, braced, plumbed and leveled, then nailed through shoe into sill beam and to each other.

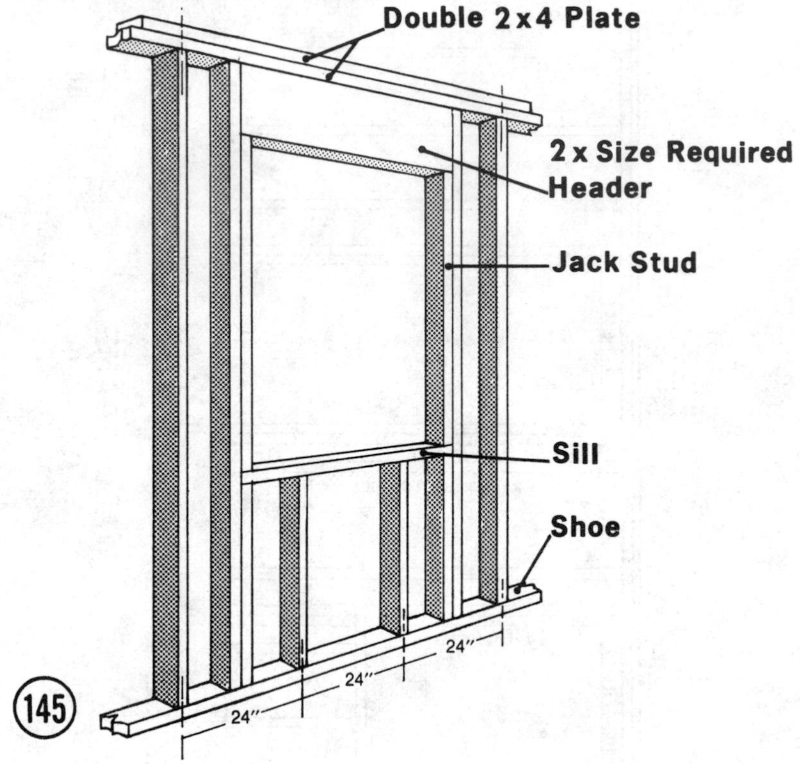

The top plate, Illus. 148, is then nailed in position. The top plate always overlaps lower plate at corners.

The ceiling joists, Illus. 149, are then toenailed to plates with 16 penny nails. These are spaced 16″ or 24″ on centers. The joists are floored over with ½, ⅝, or ¾″ plywood. If you are building a two story house, Illus. 150, you again assemble wall frames and follow procedure previously outlined.

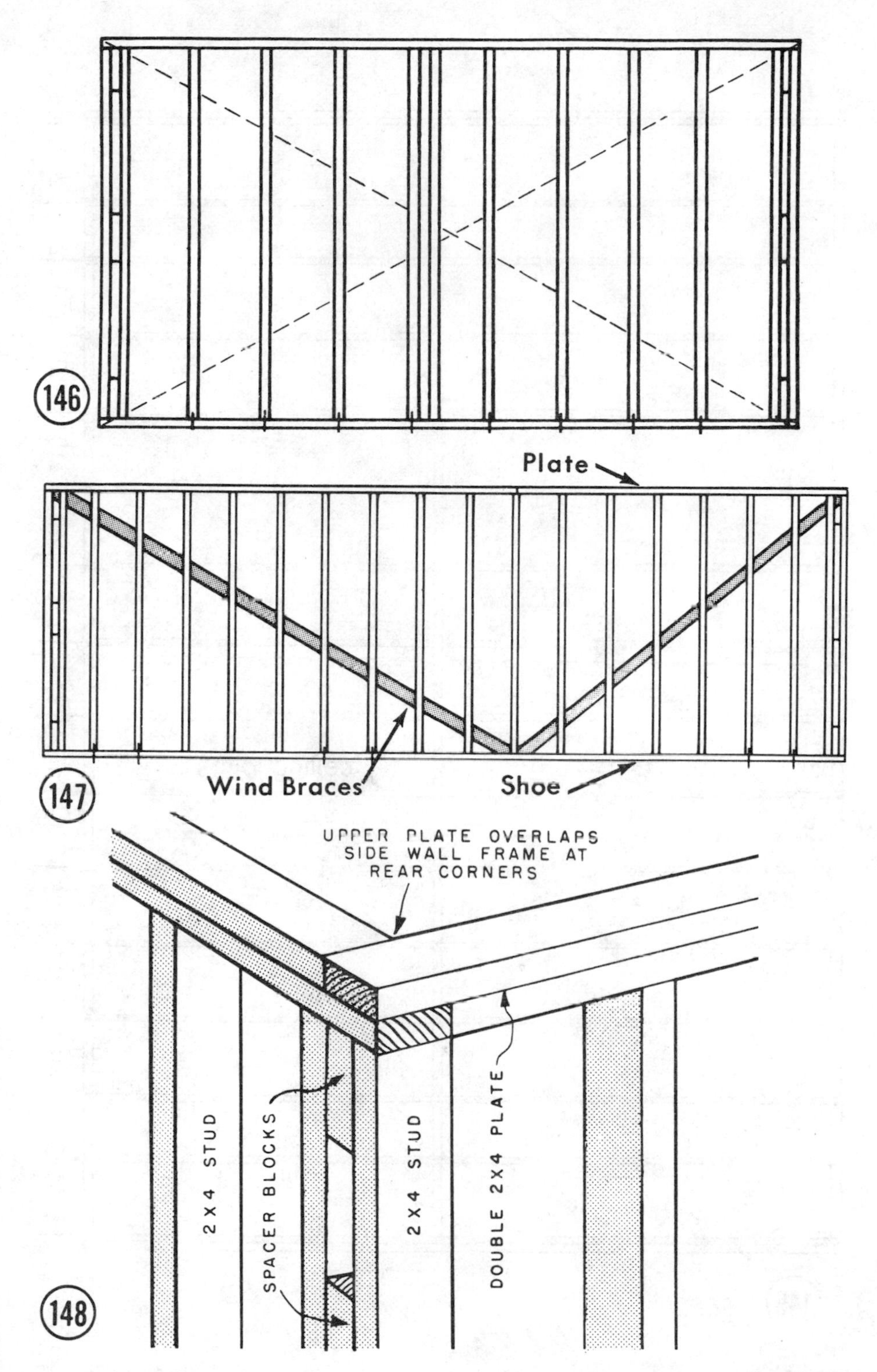
146
Plate
Wind Braces
Shoe
147
UPPER PLATE OVERLAPS
SIDE WALL FRAME AT
REAR CORNERS
2 X 4 STUD
SPACER BLOCKS
2 X 4 STUD
DOUBLE 2X4 PLATE
148

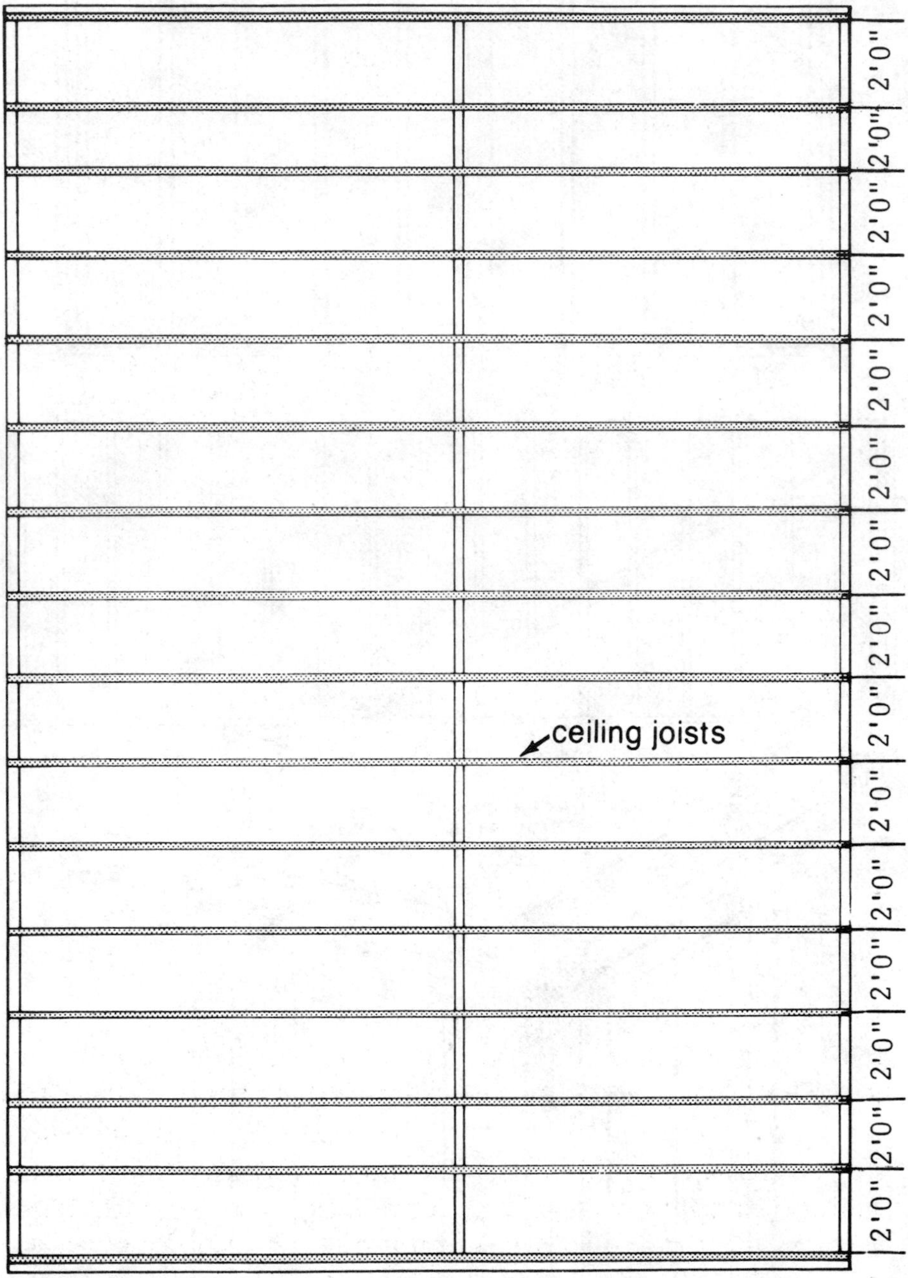

2'0"
2'0"
2'0"
2'0"
2'0"
2'0"
2'0"
2'0"
2'0"
2'0"
2'0"
2'0"
2'0"
2'0"
2'0"

ceiling joists

149

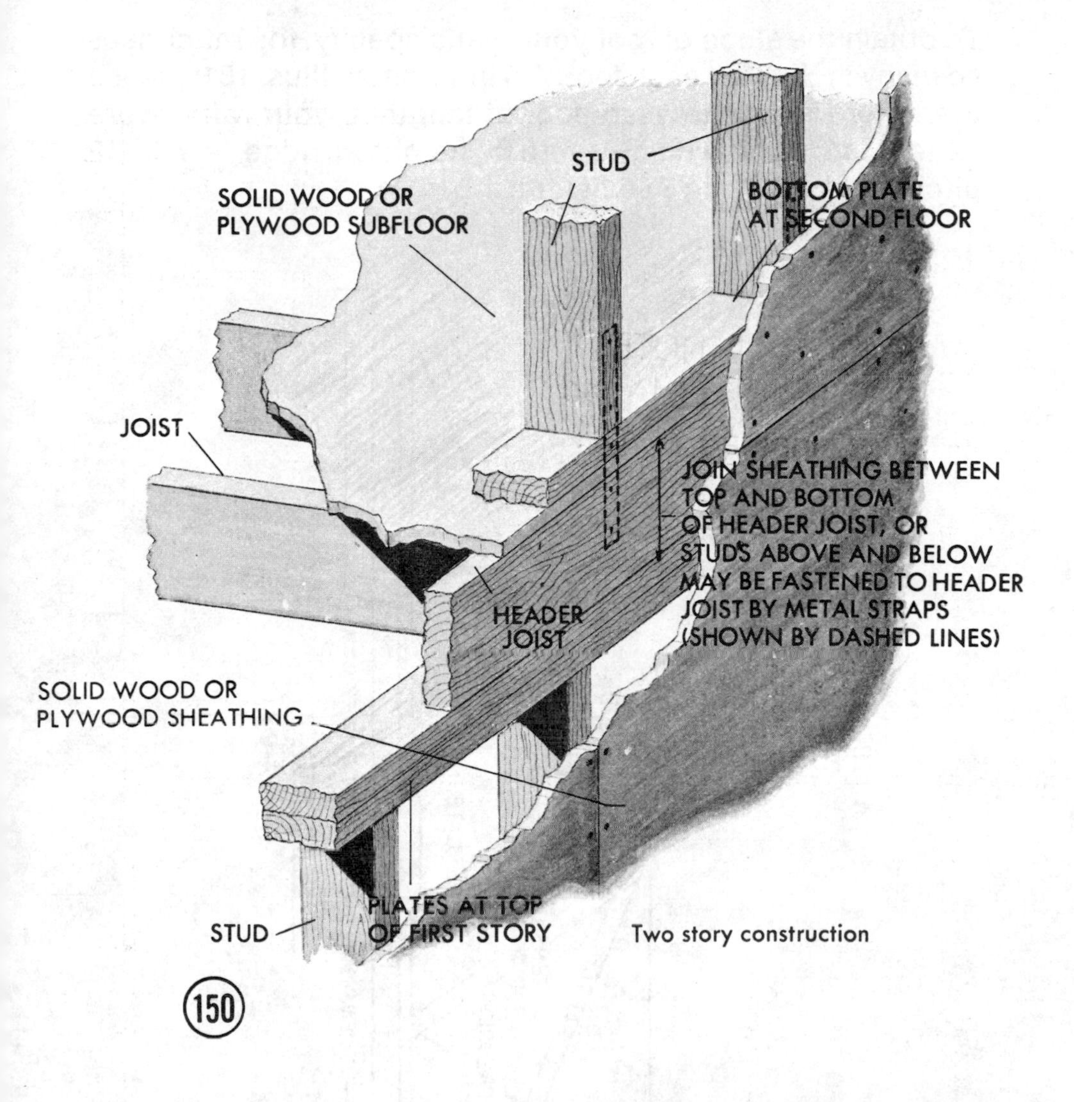

Two story construction

If you are building a one story house, the platform permits laying out rafters. Rafters are cut and tested to length and angle required, raised and braced in position with a length of ridge. Rafters can be constructed piece by piece or a prefabricated truss rafter, Illus. 151, can be purchased from most building supply centers. Most codes approve spacing truss rafters 24″ or more on centers. In this case you won't need joists and plyscord flooring in attic. 5/4 x 4, Illus. 151A, nailed 16″ on centers provide nailors for ½″ or ⅝″ gypsum ceiling panels.

To obtain the slope of roof your plans specify, the rafter rises so many inches for each foot. A 5 in 12 pitch, Illus. 151, means the rafter rises 5″ to every foot of length. If your rafter were 12 feet long, you would have 5′0″ height at ridge. A 7 in 12 pitch would provide 7′0″ height.

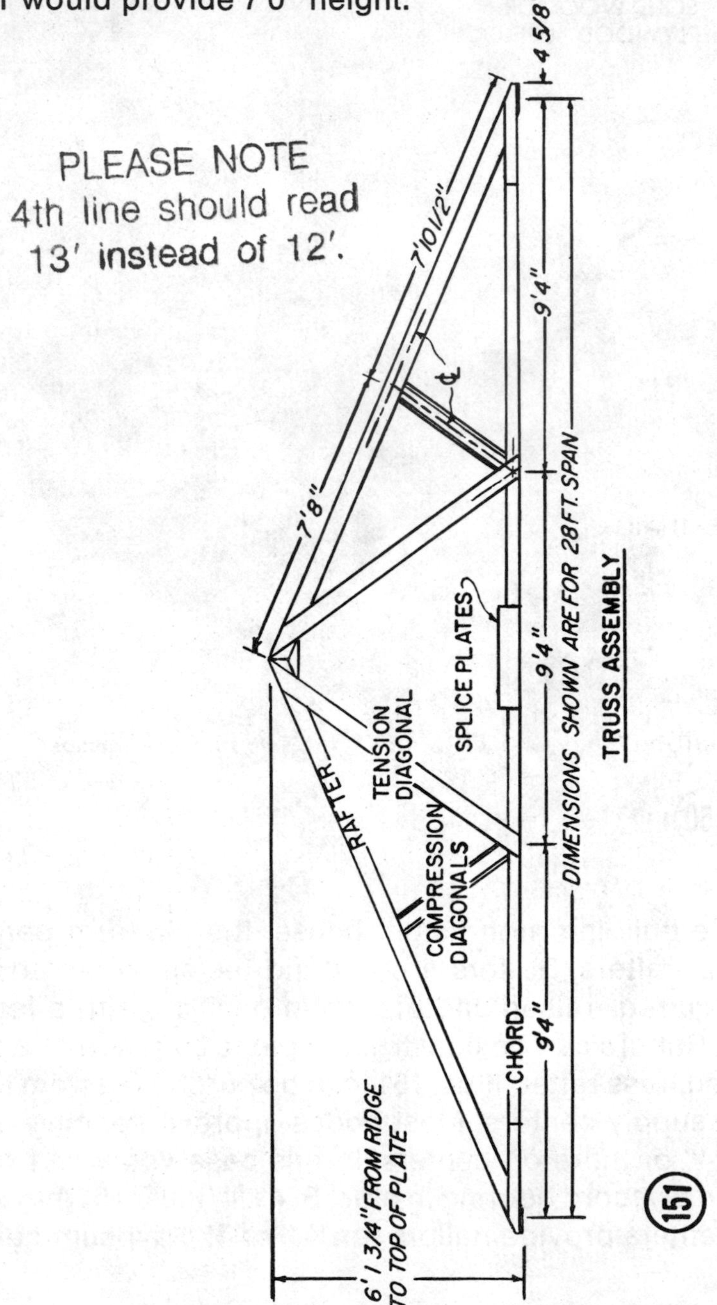

(151) a

Note Illus. 219.

Illus. 151a, shows 1 x 4 furring nailed to chord.

Building codes have established standards that dictate acceptable methods of construction. While codes in one area permit 8 x 8 x 16″ or 8 x 8 x 18″ blocks for one and two story dwellings, codes in another area require a minimum of 10 x 8 x 16″. From the earliest days of house construction carpenters spaced framing studs, joists and rafters anywhere from 16, 18, 20, 22, 24, to 26″, even 30″ on centers. Codes during the past zeroed in on 16″ spacing for all framing, Illus. 146. Step-by-step directions cover both the 16″ and 24″ module. A single 2 x 4 sill is satisfactory for size window shown, Illus. 145. Use a double 2 x 4 sill if code specifies same.

Improved methods of construction sponsored by the lumber and plywood industry have resulted in code approved methods that permit placing all framing 24″ on centers. This method not only saves material and labor but also produces better construction. The 24″ framing module is approved by the four major model codes and F.H.A. The four are the Southern Standard Building Code 1969, Basic Building Code 1970, One and Two Family Dwelling Code 1970, and the National Building Code 1967. FHA MPS 300 accepts 24″ spacing for one and two living unit buildings. Directions first explain construction using the 16″ module, then shows the 24″ module.

BEDPLATE

To locate and drill anchor bolt holes in exact position bedplate requires, it's first necessary to select and sight down the straightest lengths. Check end with a square, Illus. 152, and cut square before measuring and cutting to length required. Bedplates can be 2 x 6 or 2 x 8.

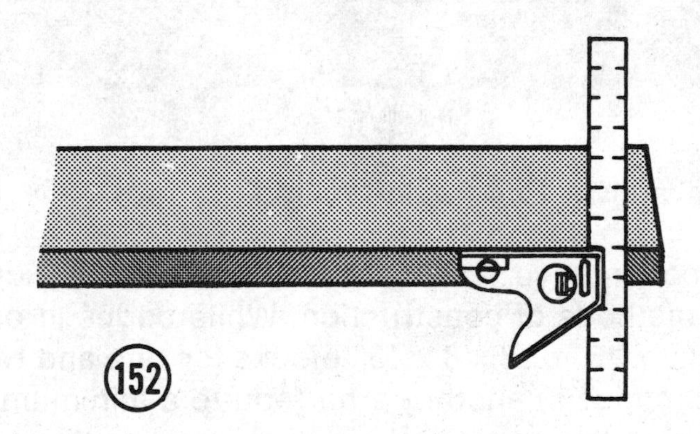

To ascertain the exact lengths, snap chalk lines on termite shield either flush with edge of foundation, or recess from edge thickness sheathing requires, i.e., 5/8″ for 5/8″ plyscord. Check chalk lines to make certain diagonals measure equal length.

After cutting a bedplate to length required, position it against anchor bolts, Illus. 153. Using a square, draw two lines to indicate exact width of bolt. Measure distance each bolt sets from edge of foundation or from sheathing, if same is recessed. Try to drill all holes in exact position required and still position bedplate on chalk line. If necessary make bolt holes slightly larger.

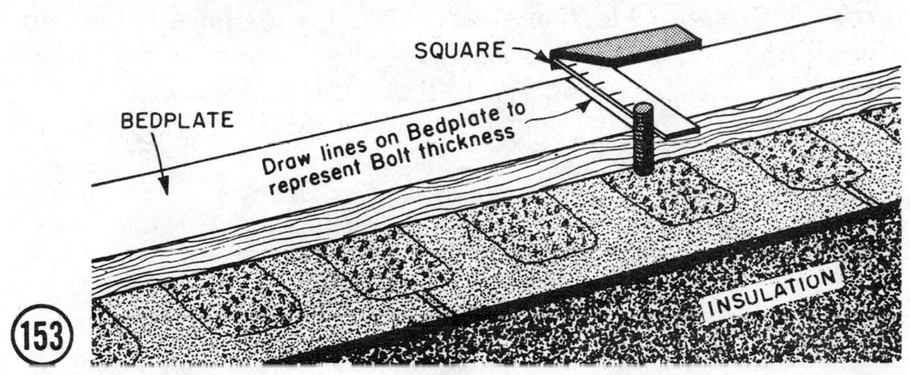

Toenail bedplates end to end. Check bedplate with a level on a 2 x 4, Illus. 10. Check spacing from bottom of 2 x 4 to bedplate with a piece of 1 x 2. Shim bedplate with pieces of shingle where needed to level. When bedplates are square and level, hand tighten nuts on anchor bolts. If you found it necessary to shim bedplates with shingle, mix one part cement to three parts of sand and flush under bedplate from both sides. Allow mortar to set three days, then tighten anchor bolts, using a wrench.

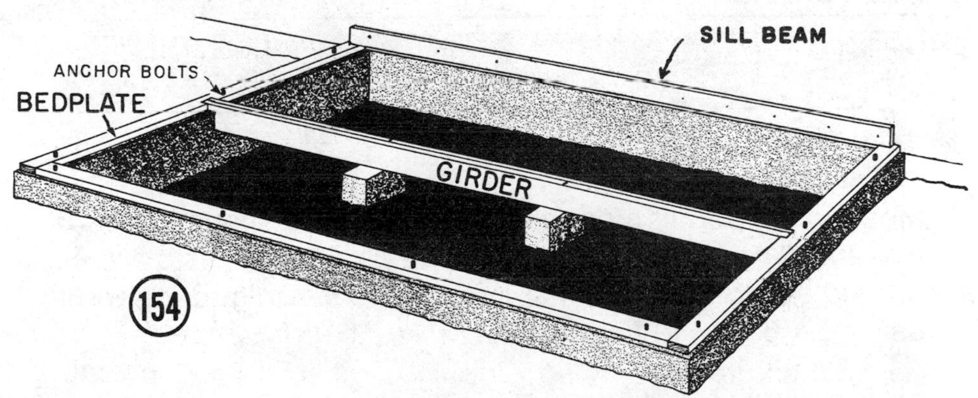

After leveling bedplate, build and install a girder, Illus. 154, if same is needed. Footings for a girder should go down below frost level. Using blocks or a round form, build piers to height girder requires. Measure width of lumber selected for a girder and build piers to height needed. Since a girder must be level with bedplate, Illus. 155, stretch and raise a guide line from bedplate to bedplate, Illus. 156. Check spacing between line and top of girder with another piece of 1 x 2. If pier is too high, plane bottom of girder. If too low, shim girder with slate. Flush mortar under girder if you need to shim it up.

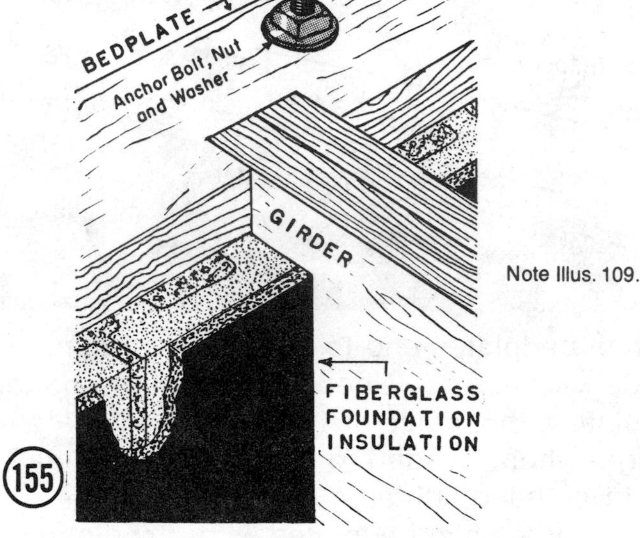

Note Illus. 109.

GIRDER

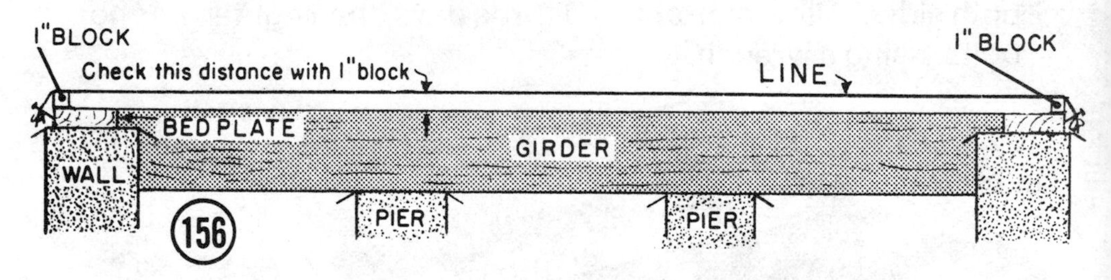

Most codes permit a double 2 x 8 girder on spans up to 10′, two 2 x 10 on spans up to 12′. Since 2 x 6 floor joists are approved by most codes for spans up to 8 ft. the girder permits using 2 x 6 x 16 ft. floor joists.

2 x 8 floor joists are approved for clear spans up to 10 feet, while 2 x 12's are approved for clear spans up to 13 feet.

NOTE: Due to the variance of lumber length and thickness, always check length of all inner framing members against your construction before cutting. Check end with a square, before measuring and cutting to length required. Always stack materials on blocks according to the size and length to speed up handling.

In specifying size, lumber is figured as follows: 1 x 2 — ¾ x 1½; 1 x 4 — ¾ x 3½; 2 x 4 — 1½ x 3½; 2 x 6 — 1½ x 5½; 2 x 8 — 1½ x 7¼; 2 x 10 — 1½ x 9¼; 2 x 12 — 1½ x 11¼.

Always sight down each piece of lumber selected for a girder and keep the "crown edge" up. Build girder using two 2 x 6, 2 x 8 or 2 x 10 or whatever size local codes specify. Stagger joints, Illus. 154, and spike together with 16 penny nails. Always butt joints over a pier.

SILL BEAMS

Use the same size lumber for a sill beam as you do for floor joists, Illus. 129. Toenail sill beam in position to bedplate. Spike sill beam to end of sill beam.

Check diagonals to make certain sill beams are square. If the bedplate was square and you nailed the sill beam in line with bedplate everything should work out O.K.

FLOOR JOISTS

Check length of floor joists required, Illus. 139, against construction. Always sight down each joist and keep the crown edge up. Using a square, measure and mark location of floor joist along top edge of sill beam.

To make certain each floor joist is level, stretch a line from sill beam to sill beam, Illus. 157. Raise line with 1 x 2 blocks. Place each joist in position and check it with a 1″ block to make certain it is the same distance from the line. Nail sill beam to floor joist with three 16 penny common nails. Toenail floor joist to bedplate with four 16 penny common nails, two to each side.

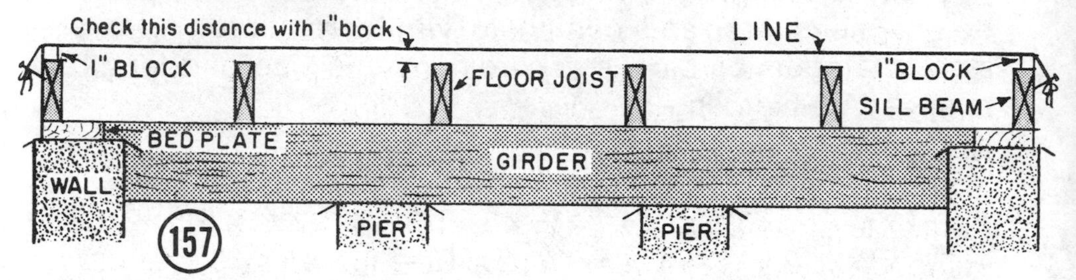

If you find that a floor joist requires shimming up or shaving off, do so before nailing in place. Toenail each floor joist to girder.

BRIDGING

Cut and nail solid bridging, using the same size lumber as the floor joist, Illus. 158. Stagger the bridging so you can drive 16 penny nails through floor joist into bridging. You can also use cross bridging, either wood or metal. In areas where winters are cold, insulate between floor joists using rock wool batts. This can be done before laying subflooring or after.

158

SUBFLOORING

Use ⅝″ or ¾″ plyscord subflooring, Illus. 142, or plyscord from your concrete forms. When you don't use adhesive, drive 8 penny nails every 8″ along edge of panel, every 12″ along intermediate joists. Keep edge flush with sill beam. Butt all joints over joists.

Illus. 143 shows the double plywood installation recommended for tile, carpet, linoleum and other other nonstructural flooring. The key to the success of this program requires the following: Use DFPA plywood underlayment. Always stagger joints in underlayment panels. Use DFPA 48/24 plywood subflooring and space the end joints 1/16″ and the edge joints ⅛″ to permit expansion and contraction. Where a single subfloor underlayment is used, Illus. 142, use bridging or blocking to support the edges of all panels. In this case, you can use 2 x 4 between joists. Always use ¾″ panels and leave 1/16″ spacing between edge and end joints.

WALL FRAMES

A wall frame consists of a 2 x 4 shoe, double 2 x 4 plate and studs spaced 16″ or 24″ on centers. Nail through shoe and single plate into each stud with two 16 penny nails. Always nail filler blocks and an extra stud in position shown to permit butting one frame against another, Illus. 159. While it's necessary to follow details provided by a building plan, if you build 8′1½″ frames, it permits using 8′ panels without cutting.

To install an opening for a window, cut and nail studs in position window requires, then nail header and sill in position. Use double 2 x dimension lumber plan specifies for a header over a window or window and door, Illus. 145. Studs are nailed to the header. Since 2″ lumber only measures 1½″, if you space headers with pieces of ½″ plyscord, header will be same thickness as the studs (3½″), Illus. 160. Nail one header in position, then nail ½″ spacers. Nail studs to second header.

Nail headers together. Spike plate to header using 16 penny nails.

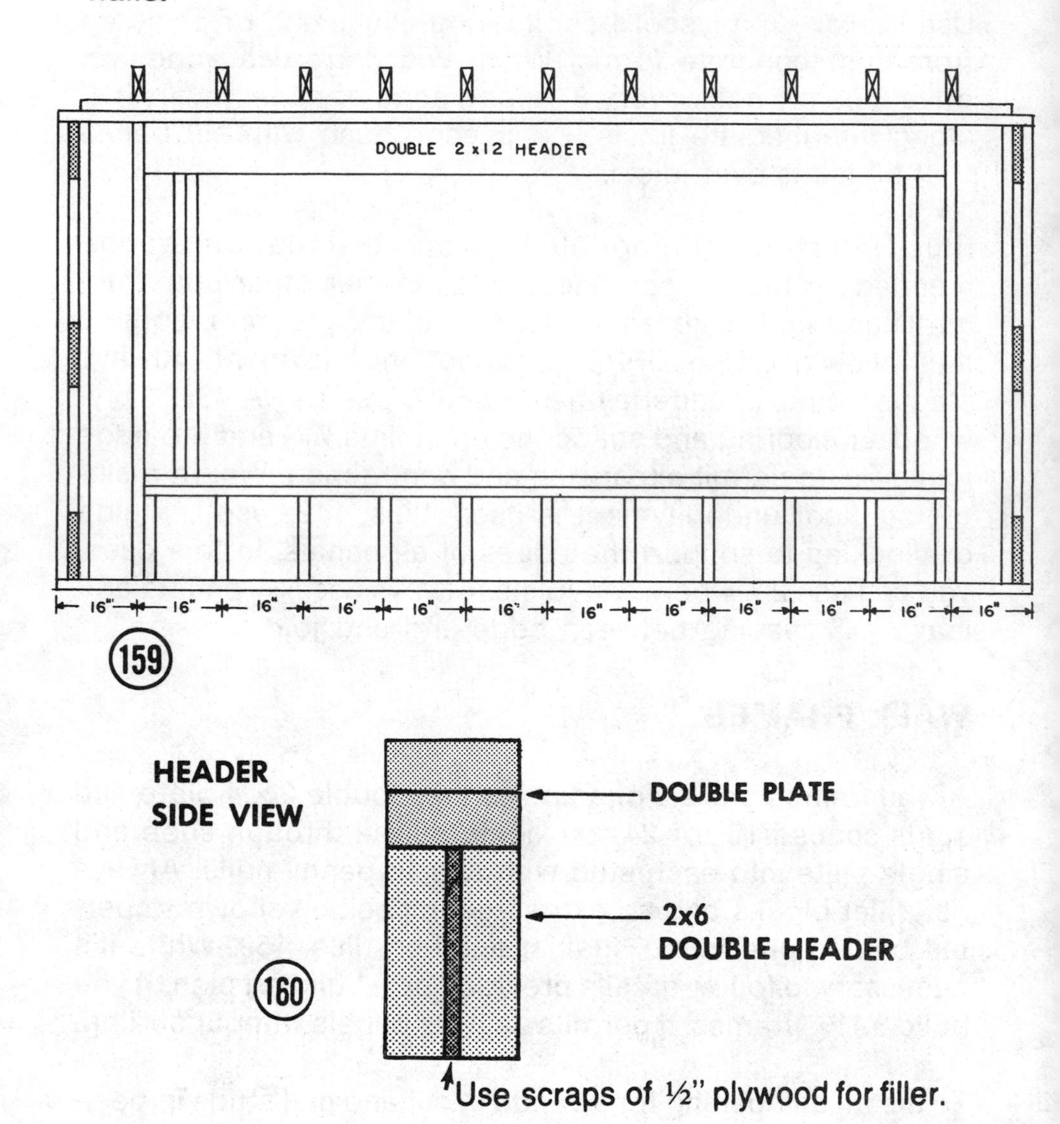

Build each wall frame on the floor and check the diagonals, Illus. 146. If diagonals are of equal length, the frame is considered square. When the frame is square, nail 1 x 4's diagonally across to brace it temporarily, Illus. 146.

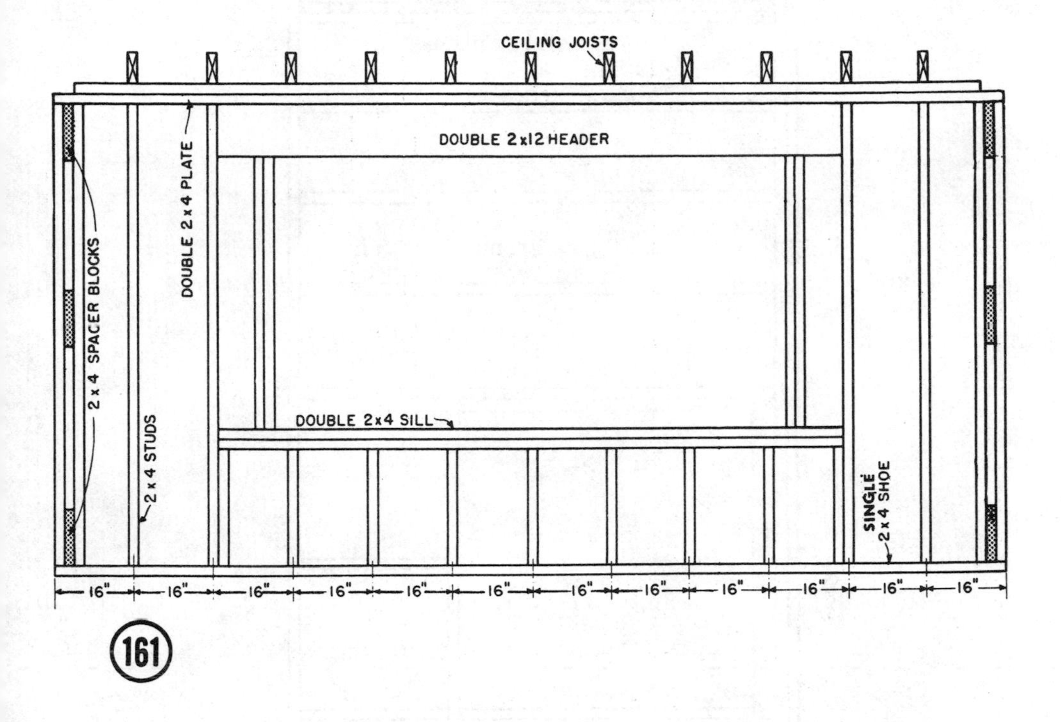

Always frame openings for a window or door to rough opening size specified by manufacturer. Always use a double header. A bow window, Illus. 159, or picture window, Illus. 161, requires double 2 x 12 headers, a double 2 x 4 sill. Most double hung and single casement windows, Illus. 162, only require a single sill.

Illus. 147 shows 1 x 4 wind braces. These are nailed in place flush with either the inside or outside edge of studs. Position 1 x 4 diagonally across frame and draw outline across each stud. Saw each stud to angle and depth brace requires and chisel out notch, Illus. 163. Cut bottom and top of brace to angle needed so it butts against stud. You can notch the plate and shoe as you do each stud, Illus. 164, or nail 2 x 4 cats flatwise, Illus. 165, ¾″ in from inside edge of studs. Nail wind braces to studs and to cats with two 8 penny nails at each joint.

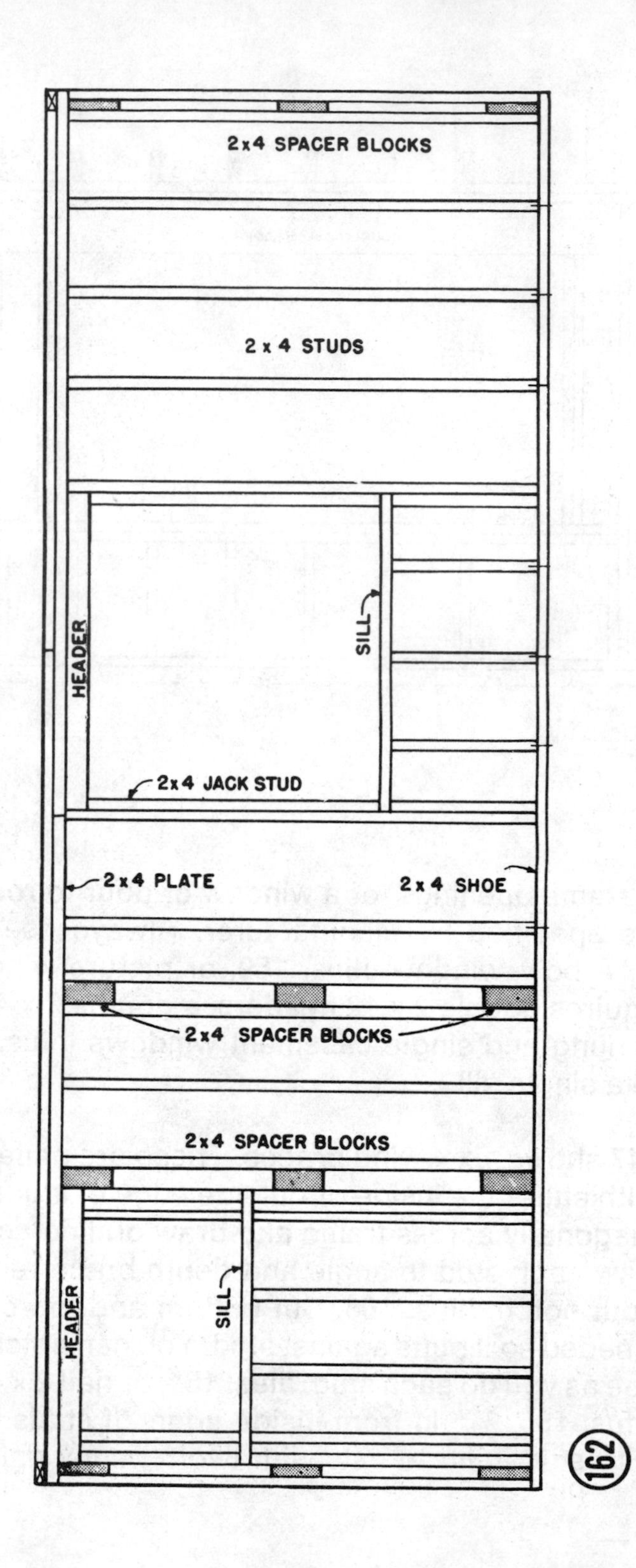
2x4 SPACER BLOCKS
2 x 4 STUDS
SILL
HEADER
2x4 JACK STUD
2x4 PLATE
2 x 4 SHOE
2x4 SPACER BLOCKS
2x4 SPACER BLOCKS
SILL
HEADER
162

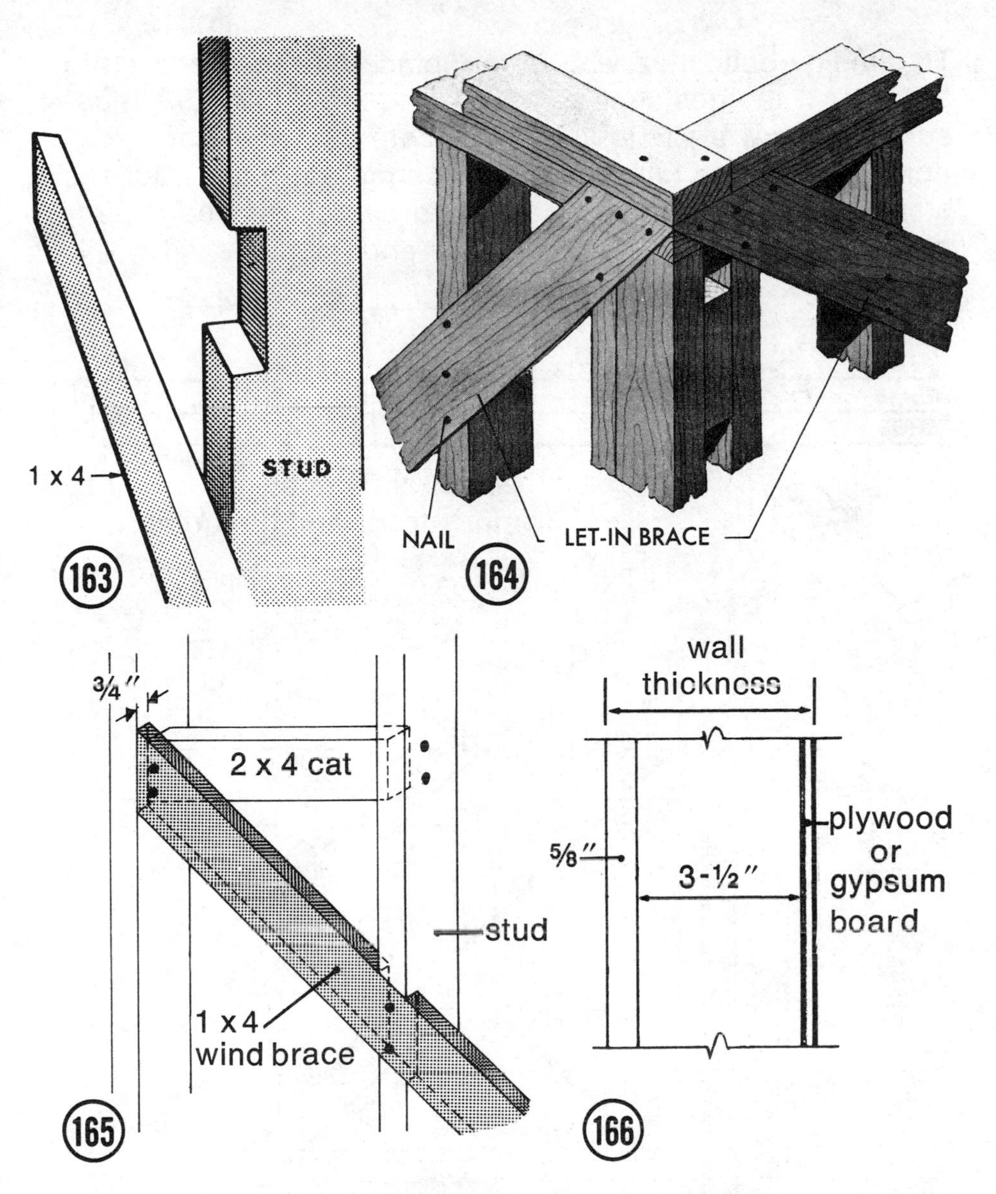

To select the size window needed for your construction, the dealer will have to know whether you are going to use plyscord subsheathing and its thickness, also whether you plan on using 1/4″ wall paneling, or apply lathe and plaster. Unless he knows the thickness of the wall he can't provide the proper depth window frame. Measure wall thickness, Illus. 166, from exterior sheathing to interior wall finish. While all codes approve 3/4″ plyscord sheathing, many approve 5/16, 3/8, 1/2 or 5/8″.

To simplify building a wall frame, place the shoe and plate together with ends square. Using a square, Illus. 167, measure and draw a single line to indicate center of each stud across both face and edge of the shoe and plate. Cut the studs to length required to maintain overall dimensions required. Nail only the lower plate to ends of studs.

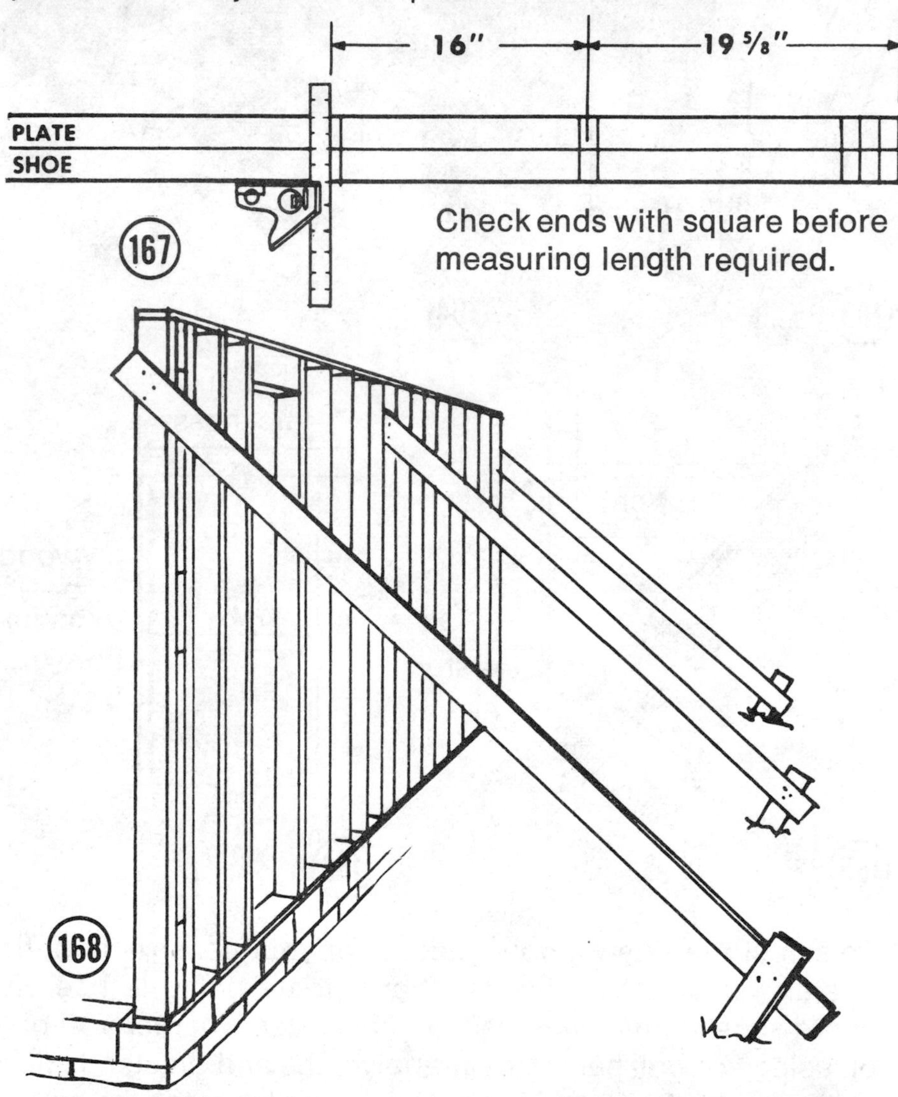

Check ends with square before measuring length required.

When using more than one length for a shoe or plate, always butt ends together under or over a stud. Nail the shoe and plate to studs with two 16 penny nails at each joint. Insert and nail spacer blocks where indicated. Any scrap piece of 2 x 4 x 8″ or longer can be used. Raise, brace and temporarily nail wall frames in position. Drive 16 penny nails through shoe into sill beam. Nail end studs together temporarily. Nail temporary braces to a long frame, Illus. 168, to 2 x 4 stakes.

Next, plumb each wall frame. Project a ruler 1″ or 2″ over plate, Illus. 169. Drop a plumb bob line over edge and measure distance from point of bob to shoe and between line and plate. When line measures same distance from plate as shoe does from point of bob the frame is considered plumb. Plumb frame in two directions at each corner.

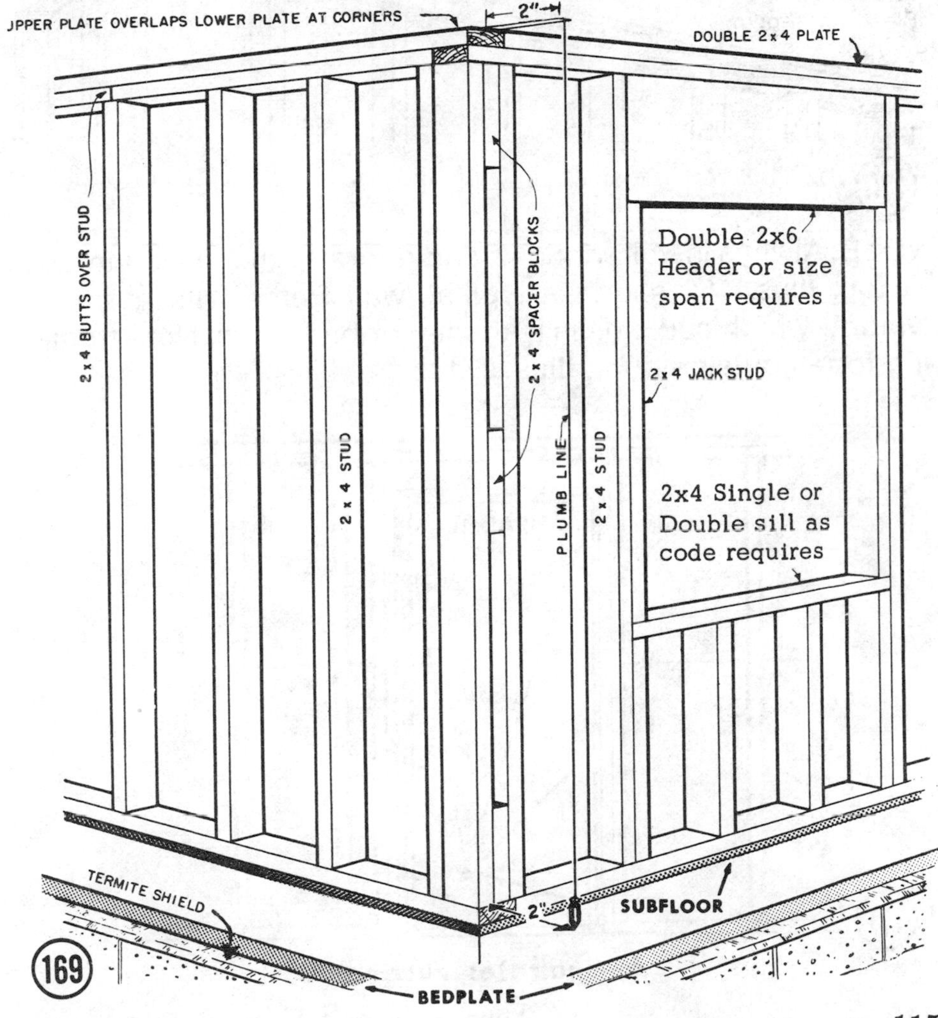

To plumb the plate, Illus. 170, nail 1 x 2 blocks in postion shown. Fasten line and check frame with another piece of 1 x 2 in a number of places. If necessary, pull frame in or push it out. Hold frame plumb with 1 x 6 braces nailed temporarily from nearest stud to a block nailed to subfloor. After frames have been plumbed, spike them together at corners. Nail the second plate in position overlapping lower plate. Illus. 148.

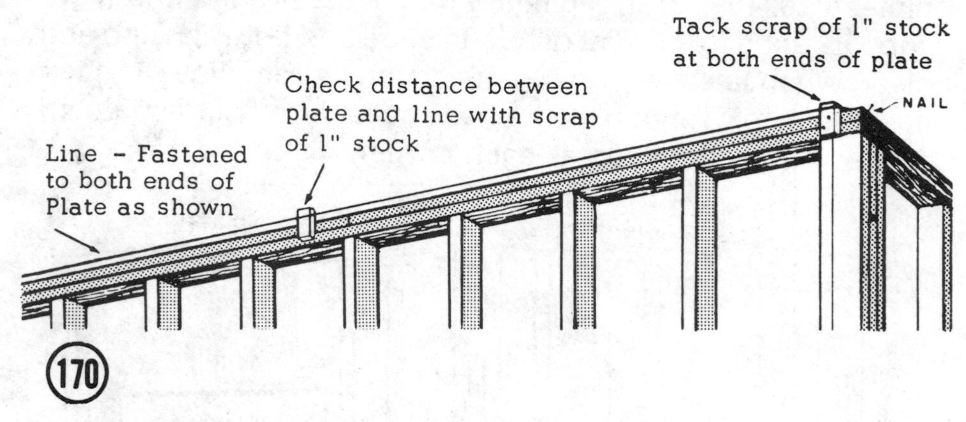

NOTE: Always install windows and doors so that they line up at top. Illus. 162 shows a typical wall frame with a small window which could be in a kitchen or bathroom. Note framing for an outside door, Illus. 171.

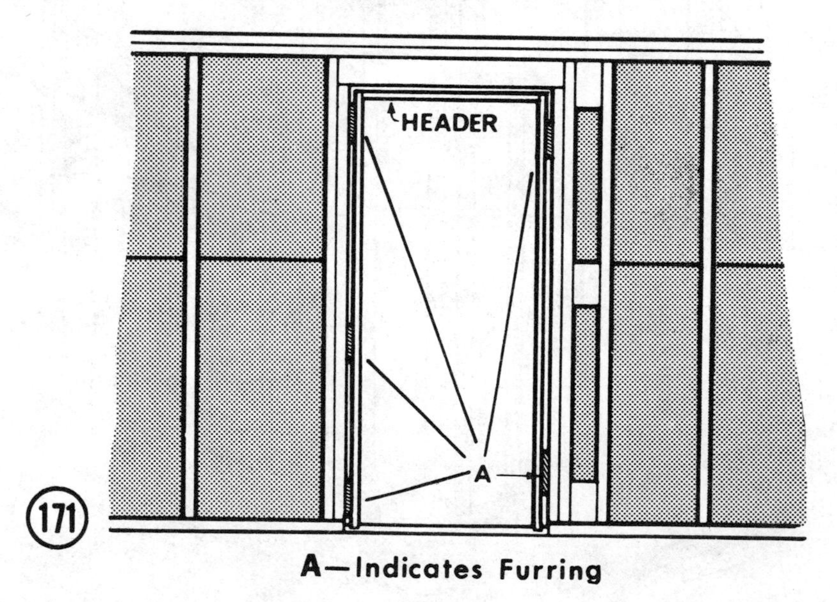

A—Indicates Furring

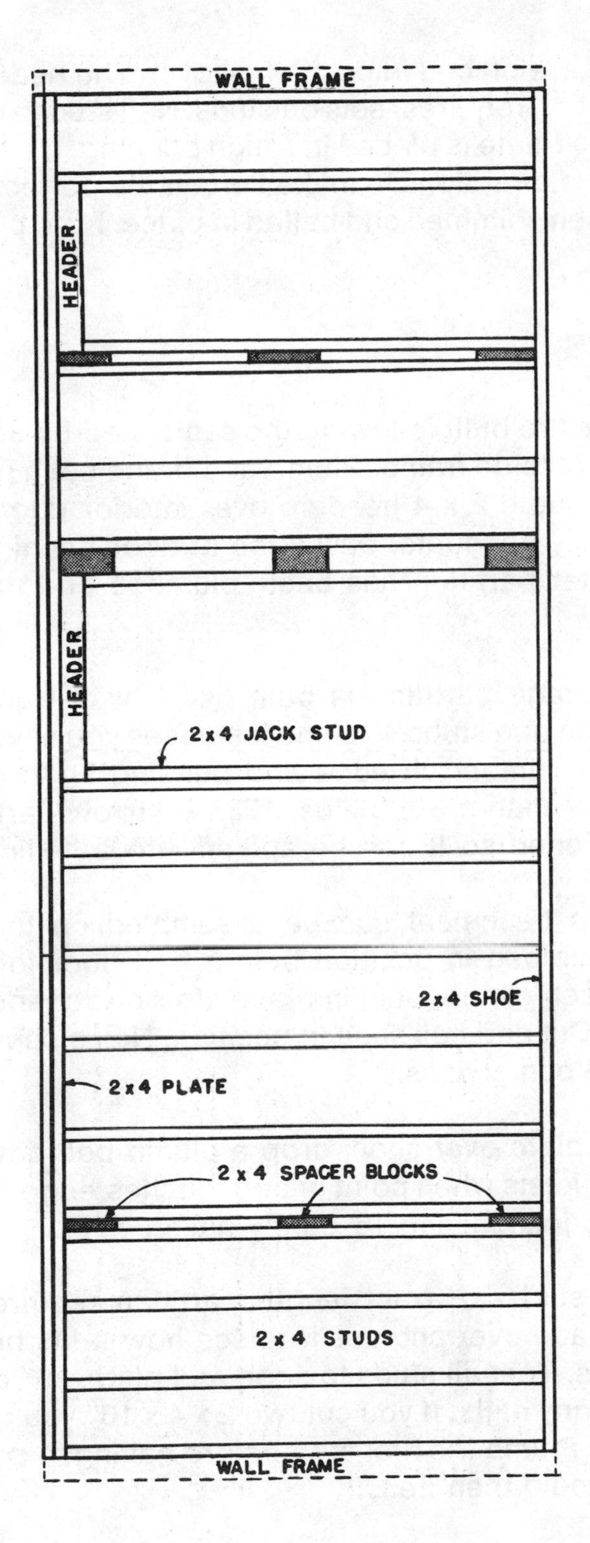
WALL FRAME
HEADER
HEADER
2 x 4 JACK STUD
2 x 4 SHOE
2 x 4 PLATE
2 x 4 SPACER BLOCKS
2 x 4 STUDS
WALL FRAME

172

Double studs are placed in position shown. The header (use size construction requires) sets on studs. Nail studs and plate to header. The frame is placed in rough opening. It should be plumbed in two directions to make certain it's perfectly level and plumb, then shimmed and nailed in place. Note blocking.

PARTITIONS

Partition walls are built following the same step-by-step procedure as an outside frame. Plumb and brace each partition in position. Double 2 x 4 headers over interior doors, Illus. 172, are usually adequate. Saw shoe at door opening flush with studs after partition has been plumbed and nailed in position.

While the average partition is built like a wall frame with 2 x 4 shoe, plate and studs, in some instances you save space by using 2 x 4 lumber on edge and building an 1½″ thick partition rather than a 3½″, Illus. 173. A narrow partition is usually positioned so it can be spiked into a ceiling joist.

Partitions in a basement can be assembled on the floor, raised and plumbed in position before first floor joists are installed or each can be built in place after girder and joists are installed. Cut and nail shoe in position. Nail sleeper clips to shoe with 8 penny nails.

To position a plate over shoe, drop a plumb bob down and draw a line on joists when point of bob touches edge of shoe. Spike plate to joists using 16 penny nails.

Cut and nail studs to exact length partition requires. Cut first stud a shade over and use it to see how it fits between shoe and plate. Toenail studs to shoe and plate, 16″ on centers, with 8 penny nails. If you cut two 2 x 4 x 16″ you can use it as a spacer. Plumb the first stud before nailing in position. The others should then be O.K.

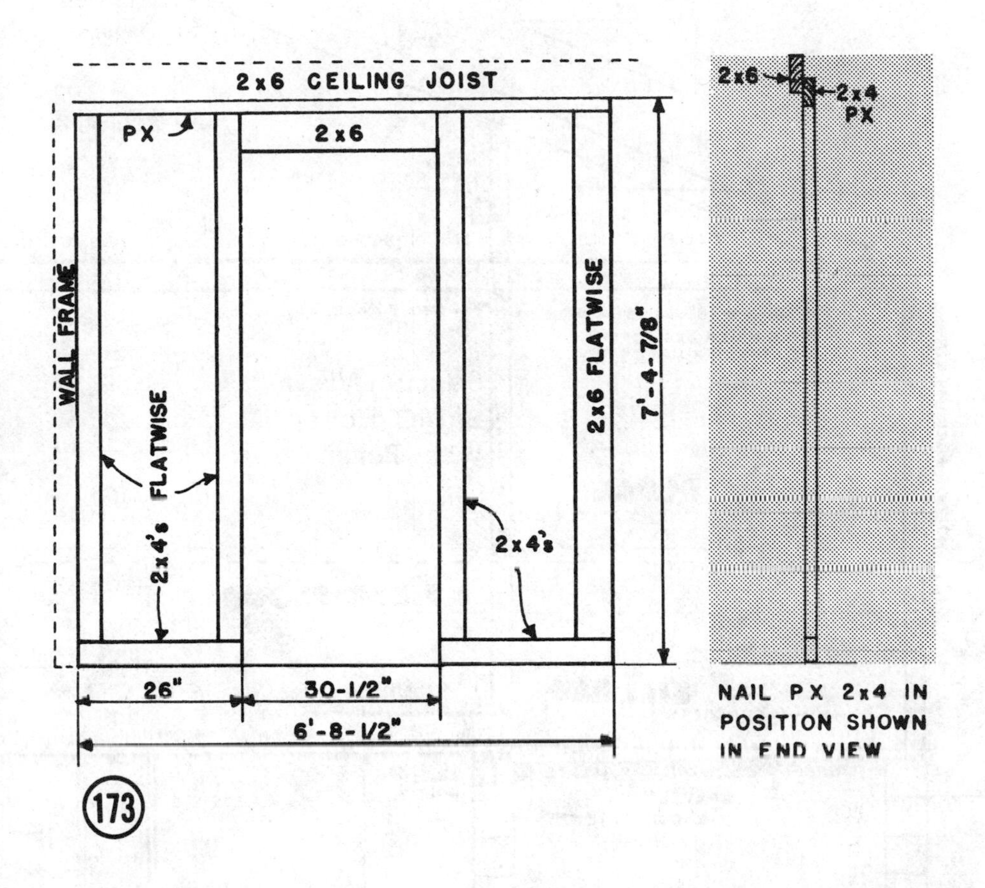

173

When building a partition in place, never cut all studs to length of the first one. Test with one, then cut to length required.

To absorb floor loading, sound construction requires use of a girder or load bearing partitions. Load bearing requires placing partitions in line, Illus. 174, or over a girder.

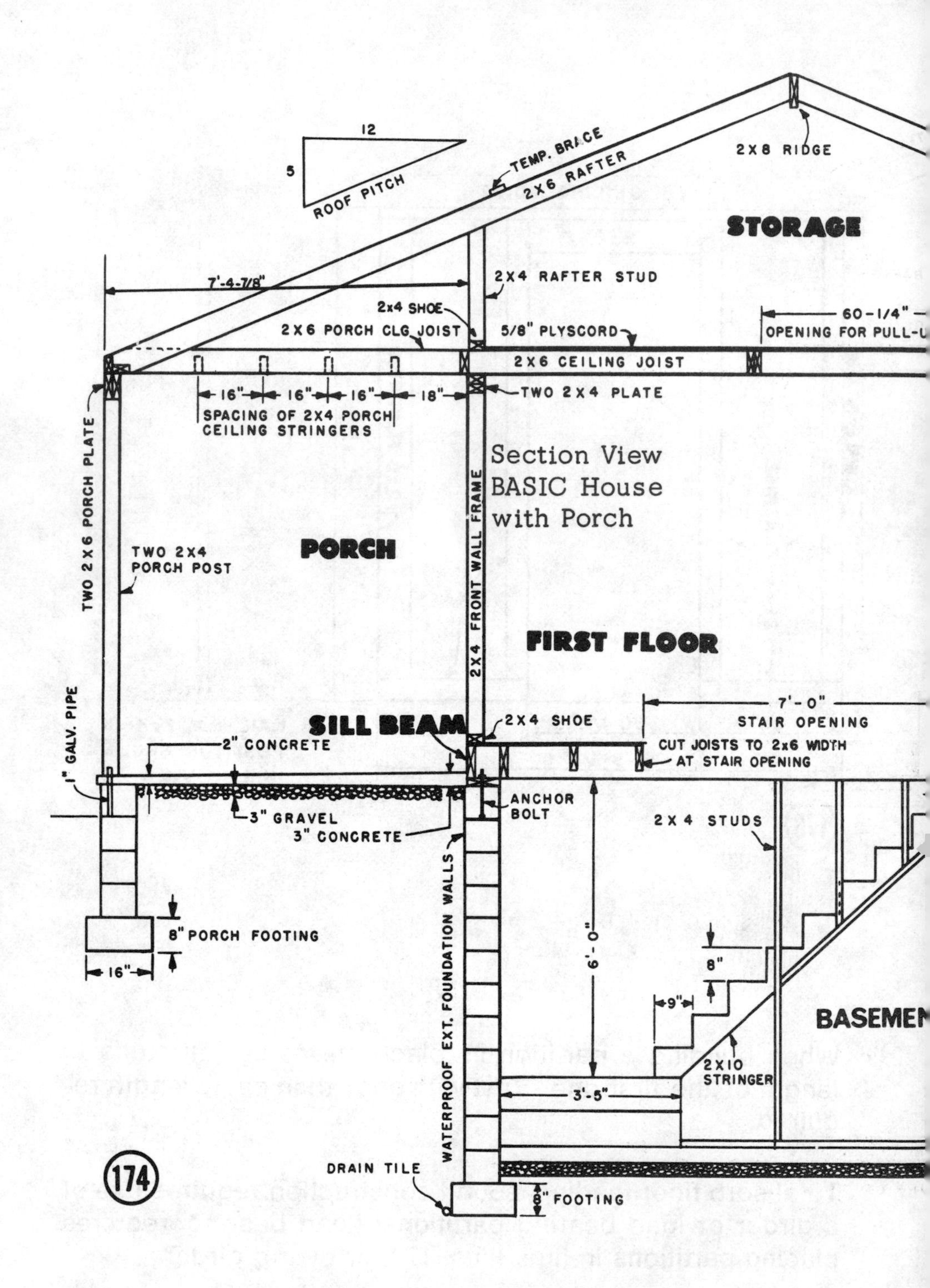

Section View
BASIC House
with Porch

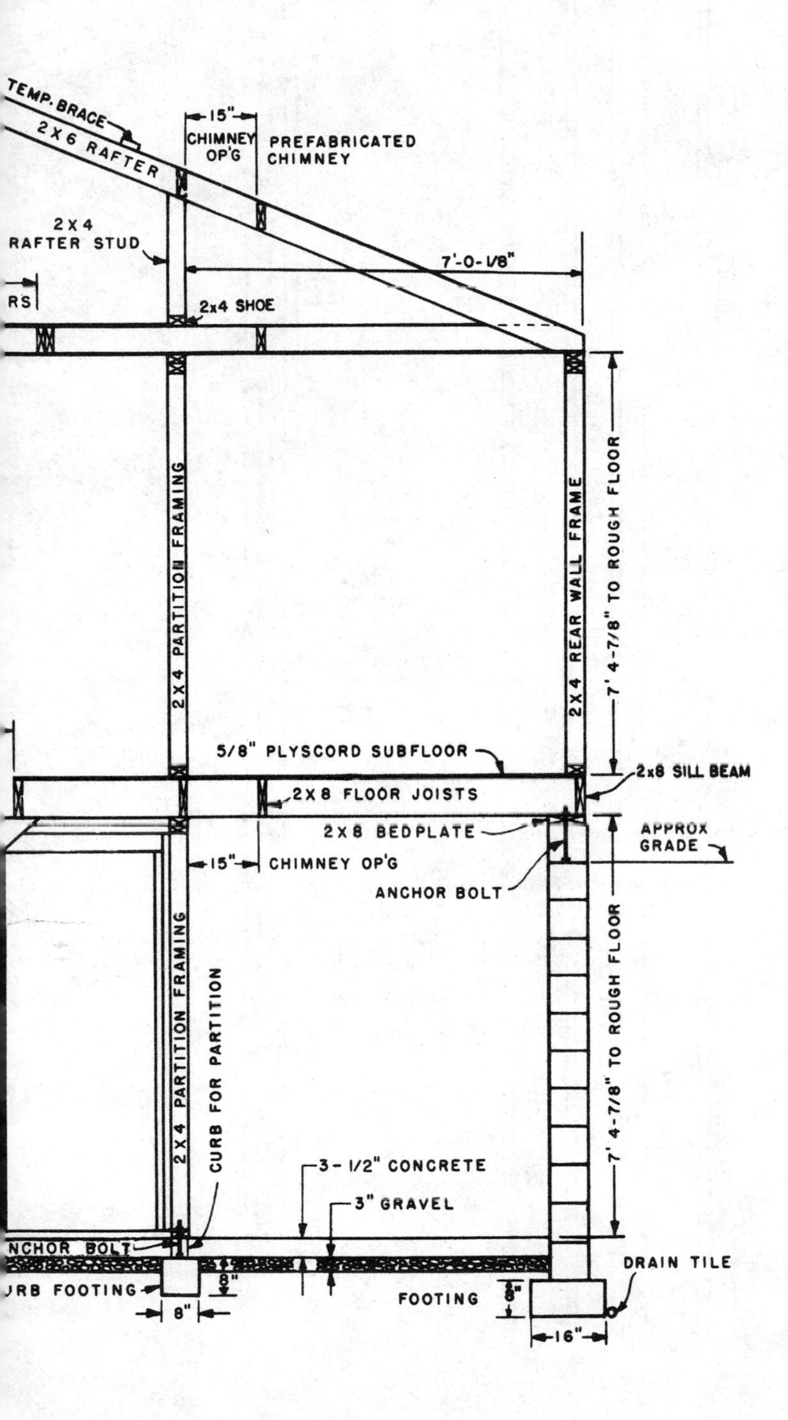
TEMP. BRACE
2 X 6 RAFTER
15"
CHIMNEY OP'G
PREFABRICATED CHIMNEY
2 X 4 RAFTER STUD
7'-0-1/8"
RS
2x4 SHOE
2 X 4 PARTITION FRAMING
2 X 4 REAR WALL FRAME
7' 4-7/8" TO ROUGH FLOOR
5/8" PLYSCORD SUBFLOOR
2x8 SILL BEAM
2 X 8 FLOOR JOISTS
2 X 8 BEDPLATE
APPROX GRADE
15"
CHIMNEY OP'G
ANCHOR BOLT
2 X 4 PARTITION FRAMING
CURB FOR PARTITION
7' 4-7/8" TO ROUGH FLOOR
3-1/2" CONCRETE
3" GRAVEL
NCHOR BOLT
DRAIN TILE
JRB FOOTING
8"
FOOTING
8"
8"
16"

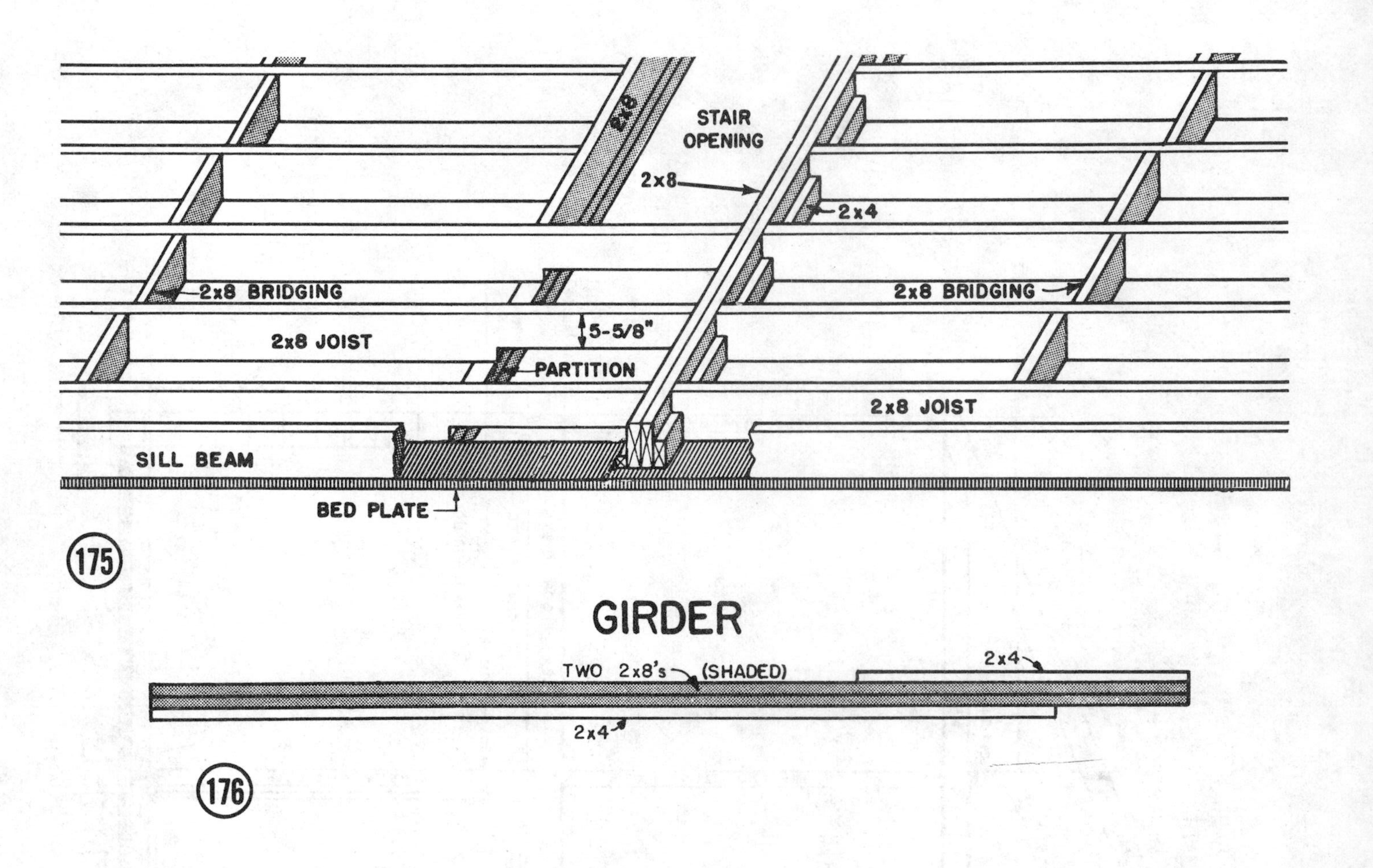
2x8
STAIR OPENING
2x8
2x4
2x8 BRIDGING
2x8 BRIDGING
2x8 JOIST
5-5/8"
PARTITION
2x8 JOIST
SILL BEAM
BED PLATE
175
GIRDER
TWO 2x8's (SHADED)
2x4
2x4
176

STAIR, CHIMNEY, FIXTURE OPENING

Illus. 175 shows framing for a stair opening. Double up and spike 2 x 8's, Illus. 176, together. Nail a 2 x 4 ledger flush with bottom edge in position shown. The double 2 x 8's acts as a supporting header. Floor joists are notched to receive 2 x 4. Toenail joists to header. Solid bridging, nailed in position shown, stiffens joists.

Use lumber equal in size to joists when installing solid bridging. Use 2 x 4 to provide an edge nailor for plyscord flooring.

Toenail each joist to bedplate and to all load bearing partitions. Always spike joists to headers. Spike joists together when they overlap over a girder or load bearing partition, Illus. 177. Joists can also butt end-to-end over a girder or load bearing partition.

Frame opening for a chimney with double headers as shown, Illus. 177. Use a plumb bob to locate and mark opening. Use same size lumber when framing headers between joists or rafters.

Spike rafters to ceiling joists with 16 penny nails, Illus. 178. Chop end of joist to angle of rafter. To hold rafters 16″ on centers, nail a 1 x 4 temporary brace across rafters, Illus. 227.

Illus. 179 shows 2 x 8 cats nailed between joists to provide a nailor for wood panels, not required for sheetrock.

If you want to recess lighting fixtures, Illus. 180, frame in rough opening to size fixtures require.

Nail or glue plyscord panels to joists following procedure previously outlined. This provides a platform to lay out rafters, raise truss rafters, or layout, assemble and raise second story wall frames.

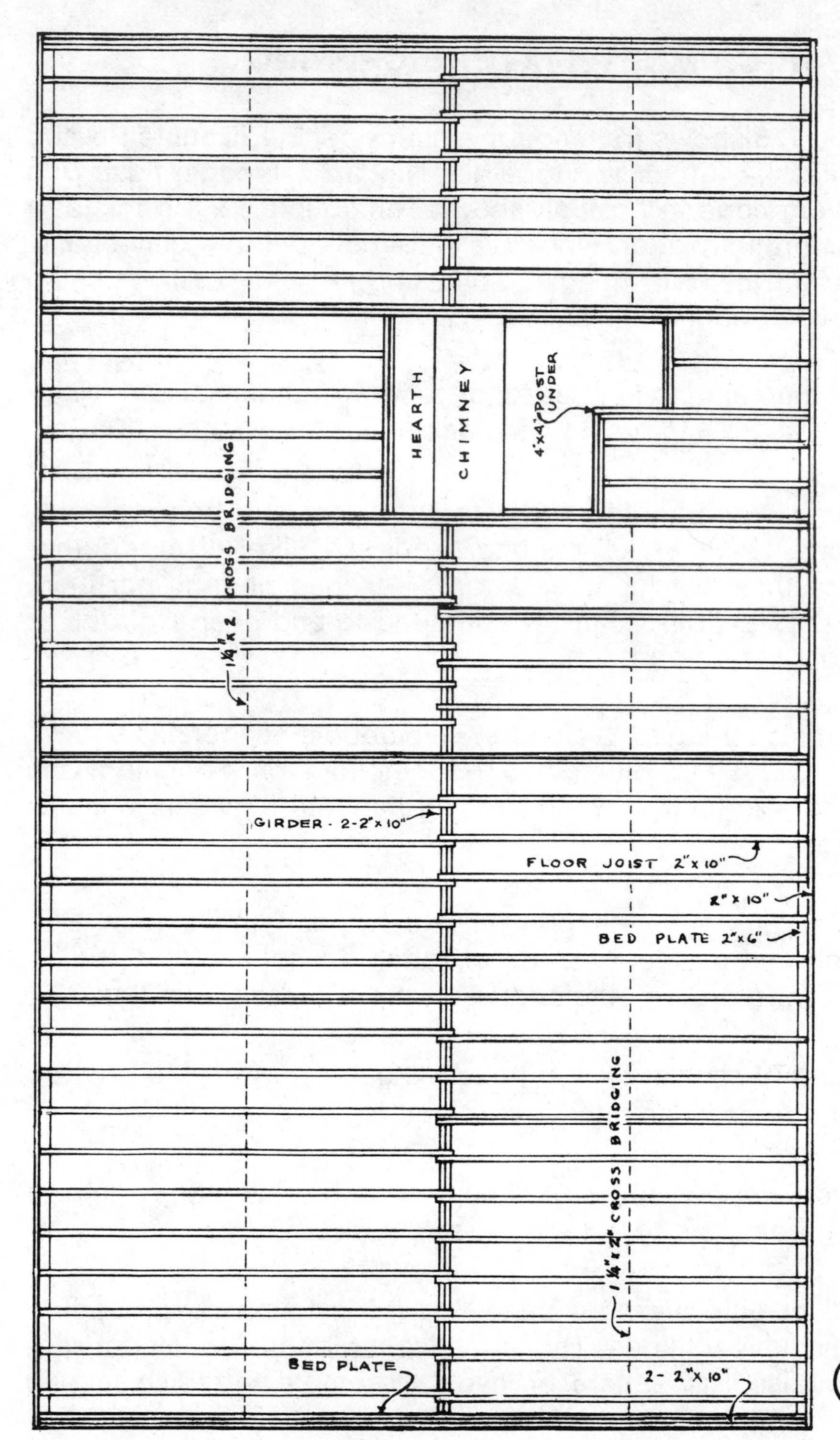
HEARTH
CHIMNEY
4'x4" POST UNDER
1¼"x2 CROSS BRIDGING
GIRDER - 2-2"x10"
FLOOR JOIST 2"x10"
2"x10"
BED PLATE 2"x6"
1¼"x2" CROSS BRIDGING
BED PLATE
2- 2"x10"

177

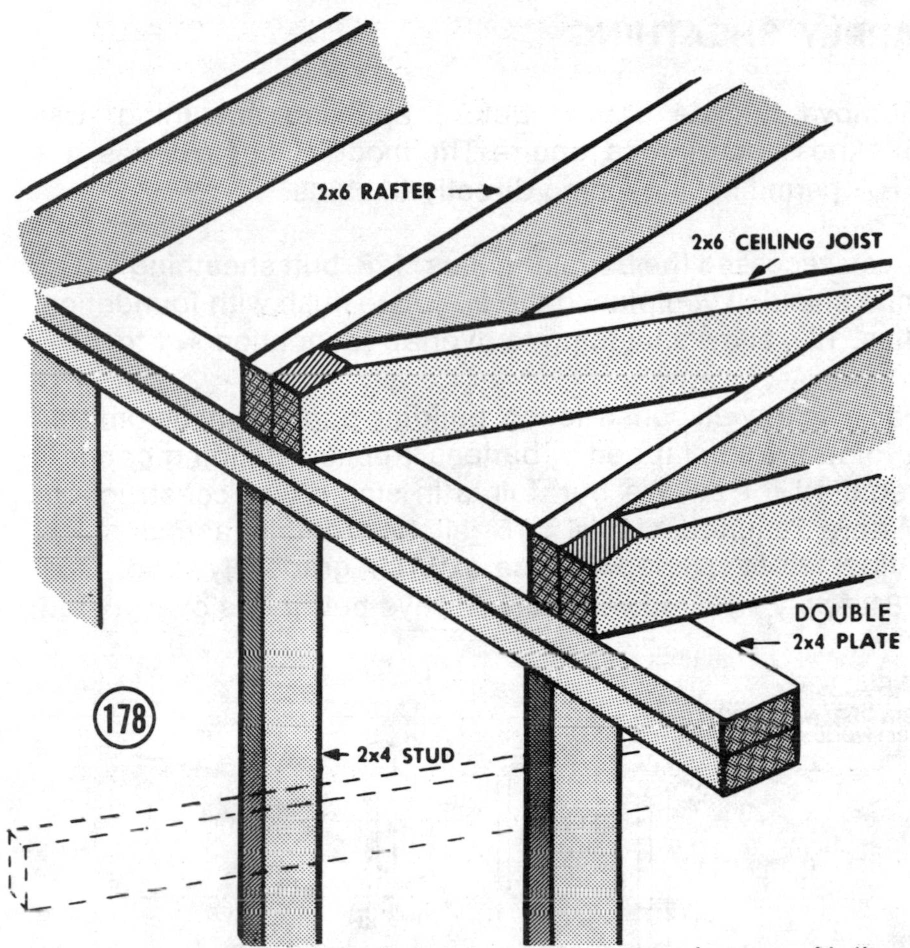

Dash lines indicate 2x4x16 temporary cross braces. Nail to studs every 4 feet.

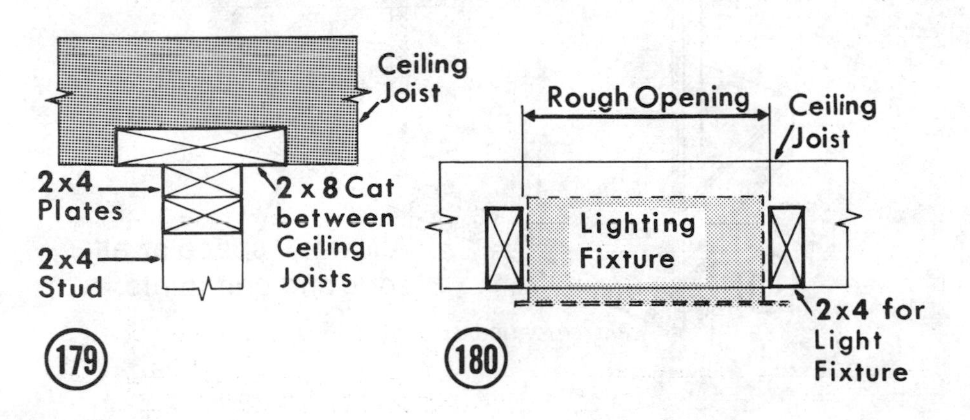

APPLY SHEATHING

Remove outside braces before applying sheathing. Use thickness local codes require. The model building codes and FHA permit nailing siding directly to studs.

If you recessed the bedplate, Illus. 128, butt sheathing to termite shield. If you placed the bedplate flush with foundation, Illus. 129, allow sheathing to overlap foundation ¾″ to 1″.

For single-wall construction, building paper may be omitted if joints are shiplapped or battens applied. Building paper is required for square butt joints in single wall construction. Where required, staple #15 felt horizontally across studs. Allow felt to overlap 4″. Use exterior grade plywood, Illus. 181. Apply plywood vertically. Always butt joints over a stud.

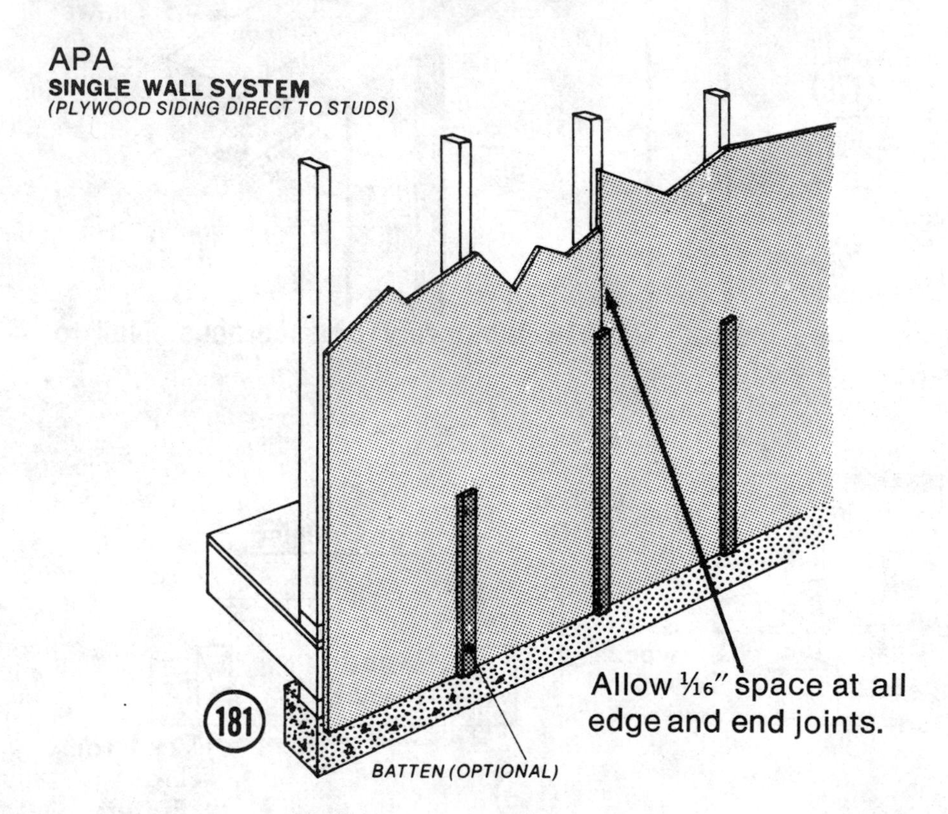

DOUBLE WALL
(SHEATHING AND SIDING)

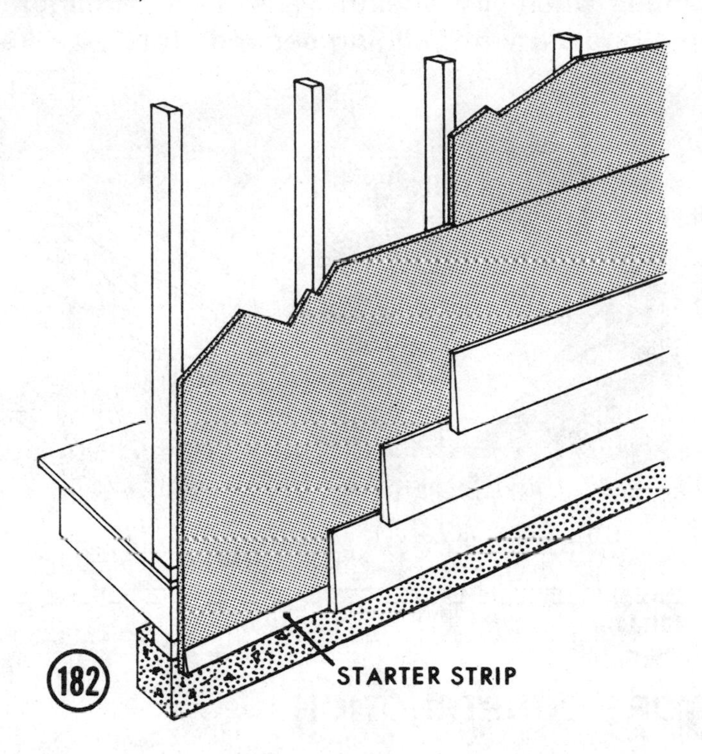

(182)

Double wall sheathing consists of 5/16, 3/8, 1/2 or 5/8″ C-D or C-C plywood sheathing plus siding. Siding may be plywood clapboard or shingles. Apply sheathing to studs vertically or horizontally, as shown in Illus. 182. Always stagger joints so ends of adjoining panels never meet over same stud, Illus. 183.

C-C grade — C grade veneer on face and back.
C-D ″ — C ″ on face, D grade on back.

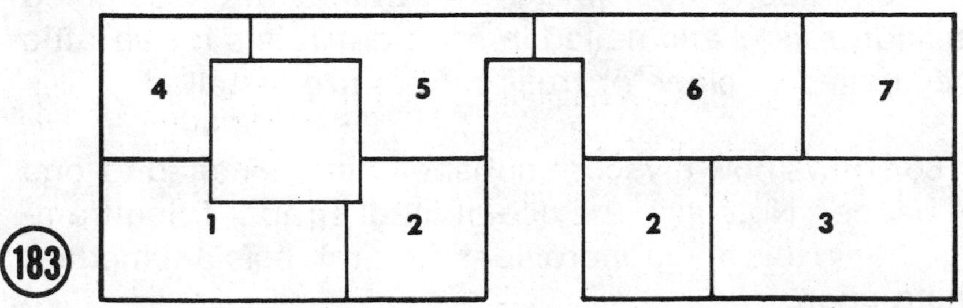

(183)

Apply wall sheathing up to plate. Illus. 184 shows method of nailing recommended for storm resistant construction. You stiffen framing when you sheath walls. This permits raising rafters on a one story or building second story wall frames.

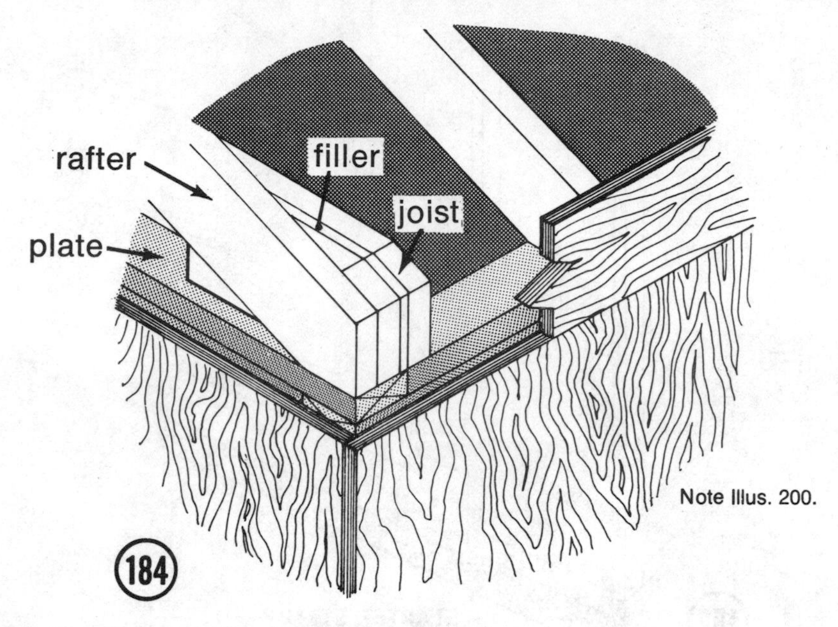

TWO STORY CONSTRUCTION

There are several approved methods of framing a second story. One method specifies nailing blocking, Illus. 185, between ends of joists. ⅝ or ¾" plyscord flooring is then nailed or glued to joists. Wall frames are assembled raised and secured in place following procedure outlined for first story.

Another method, Illus. 186, specifies a header. This is toenailed to plate. The header is nailed to floor joists. Plyscord panels are nailed to floor joists, wall frames are assembled on the floor, raised and nailed in position. Joists for an attic are then nailed in place or truss rafters are installed.

Illus. 186 shows how plyscord subsheathing is nailed to one half of header. Note also the use of steel straps. Tying framing together in this manner provides a storm resistant method of construction.

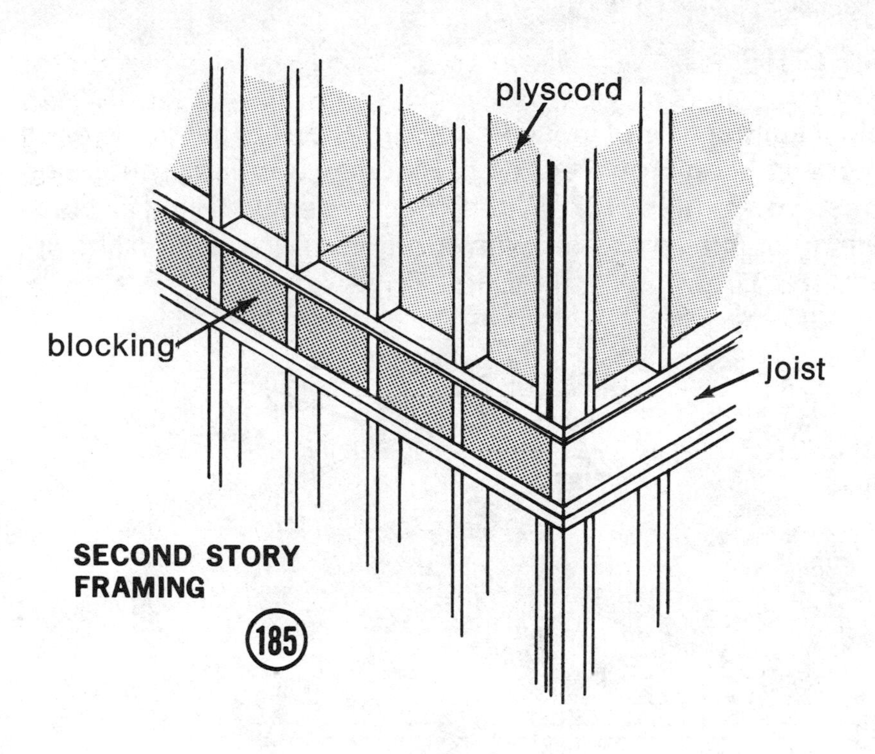

SECOND STORY FRAMING

185

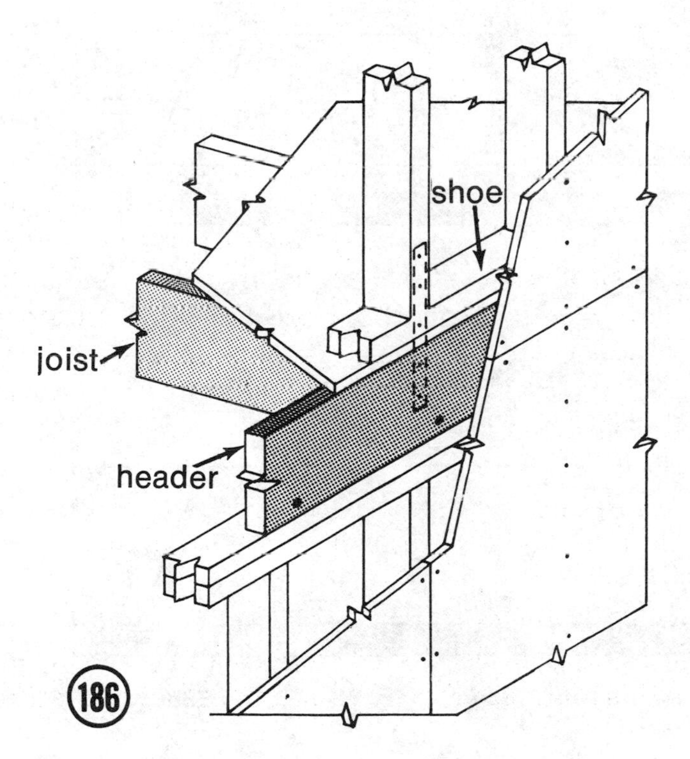

186

RAFTERS

Select lumber free of loose knots for rafters. Sight down each rafter and keep crown edge up. Place an X on edge that indicates crown, Illus. 187. When plans specify 5 in 12, place framing square on rafter with the 5 in 12 in position shown, Illus. 188. Draw line A, Illus. 189.

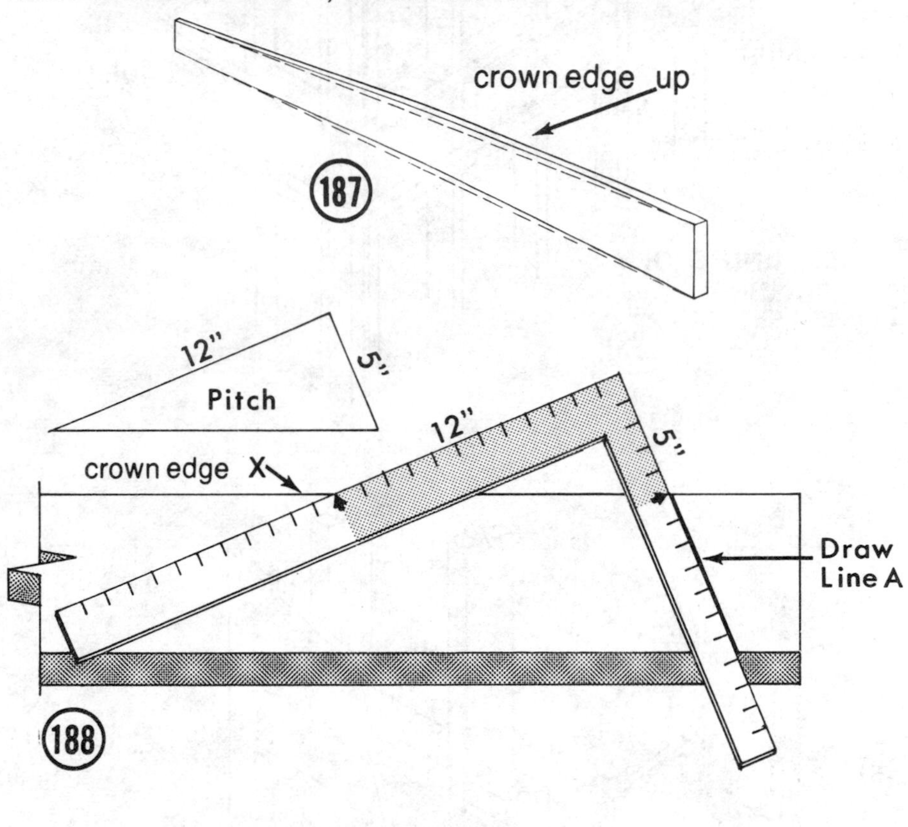

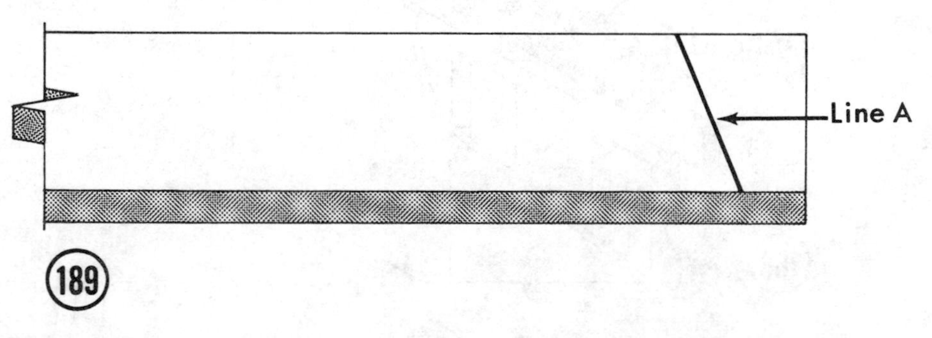

To draw angle for plate end of rafter, place square in position shown, Illus. 190. When distance B measures 3½″, draw line B, Illus. 191. Saw end of rafter to shape shown, Illus. 192.

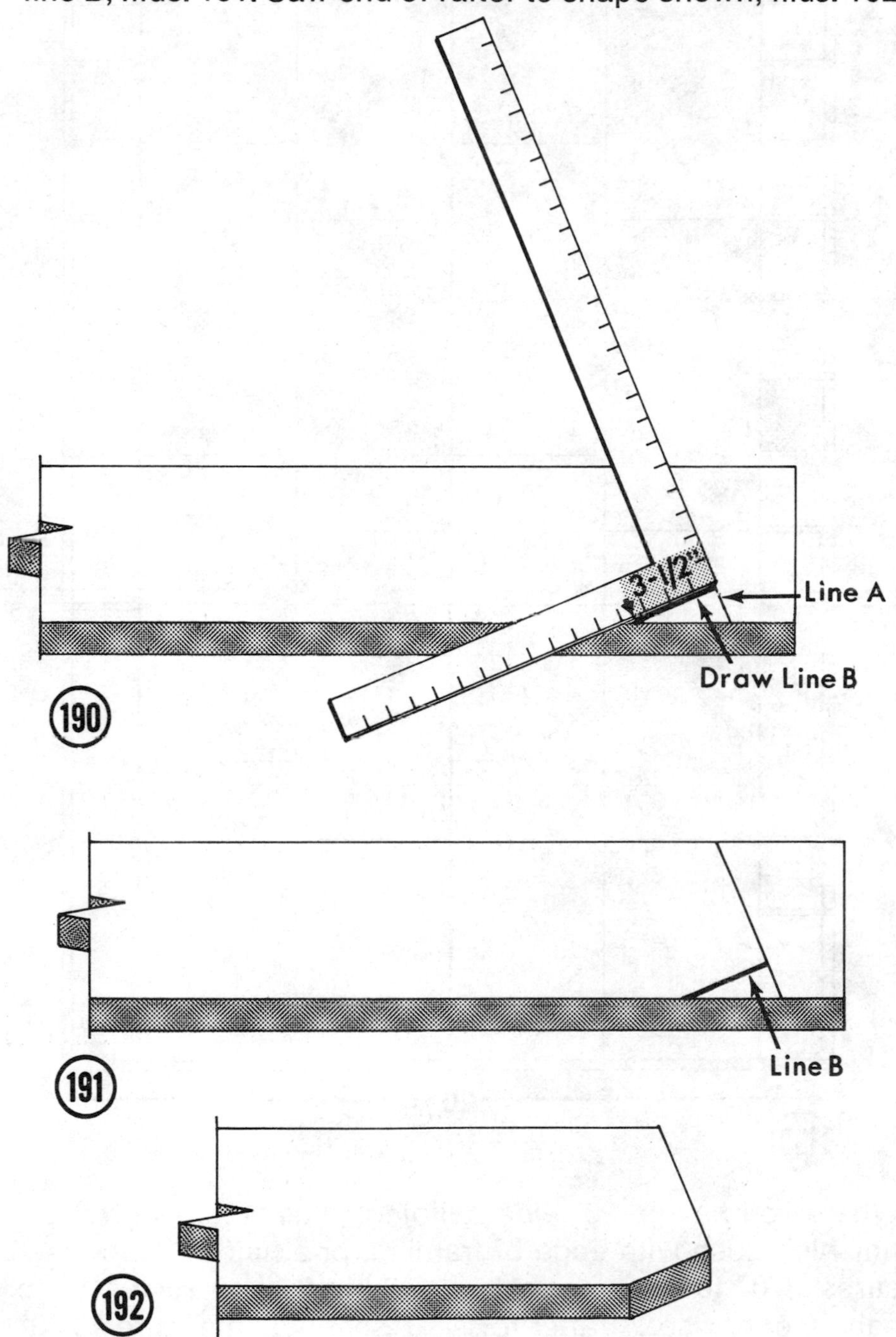

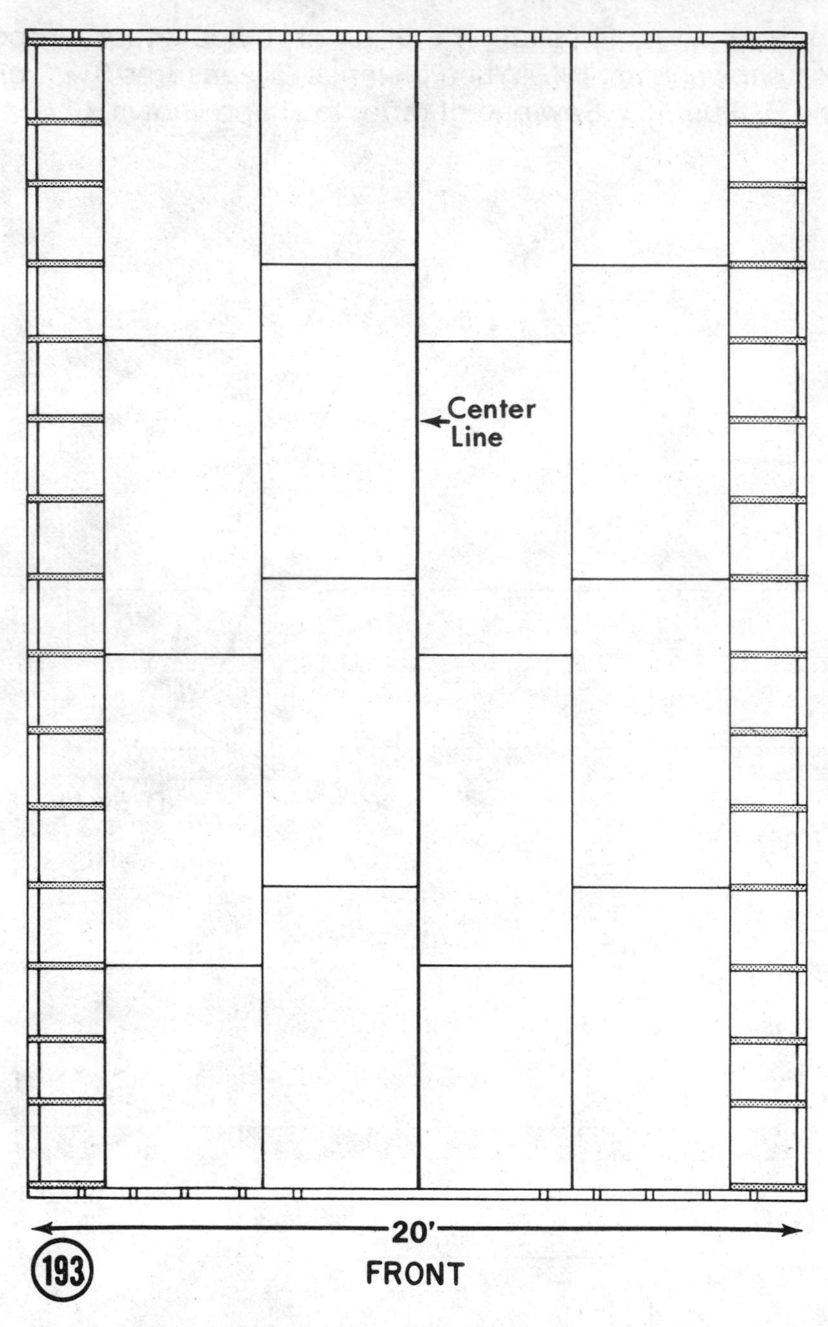

Where you have a 20′ wide building, Illus. 193, with rafters finishing flush with edge of framing, one half of span measures 10′0″ to center of ridge. A 5 in 12 pitch runs 1′1″ to each foot of rise. A rafter for a 10′ span will measure 10′10″

less half the thickness of the ridge. Measure along top edge and cut a test pair 10′10″. Using the square with 5 in 12, Illus. 194, draw angle for ridge. Cut a pair of test rafters using 1 x 6 — 10′10″ each.

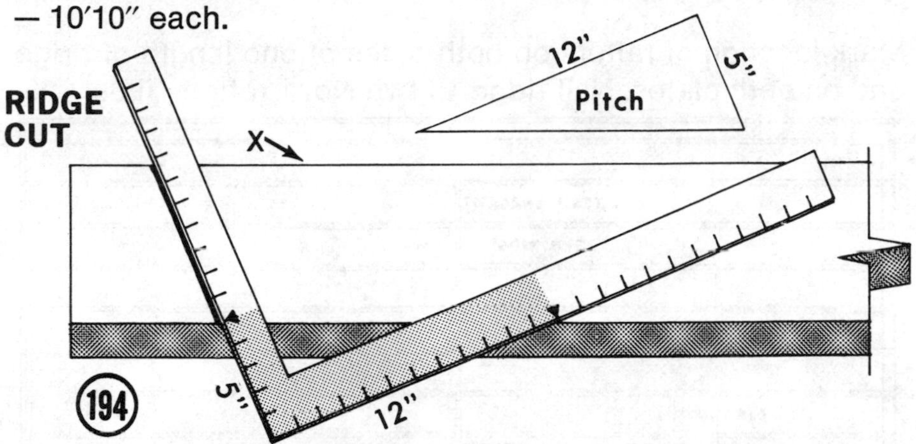

Nail 2 x 8 blocks to ends of joists, Illus. 195. Place a pair of rafters on floor. Cut pair to exact length needed when a scrap piece of 2 x 8 is used as a ridge. When the test pair proves O.K. cut others to exact length required.

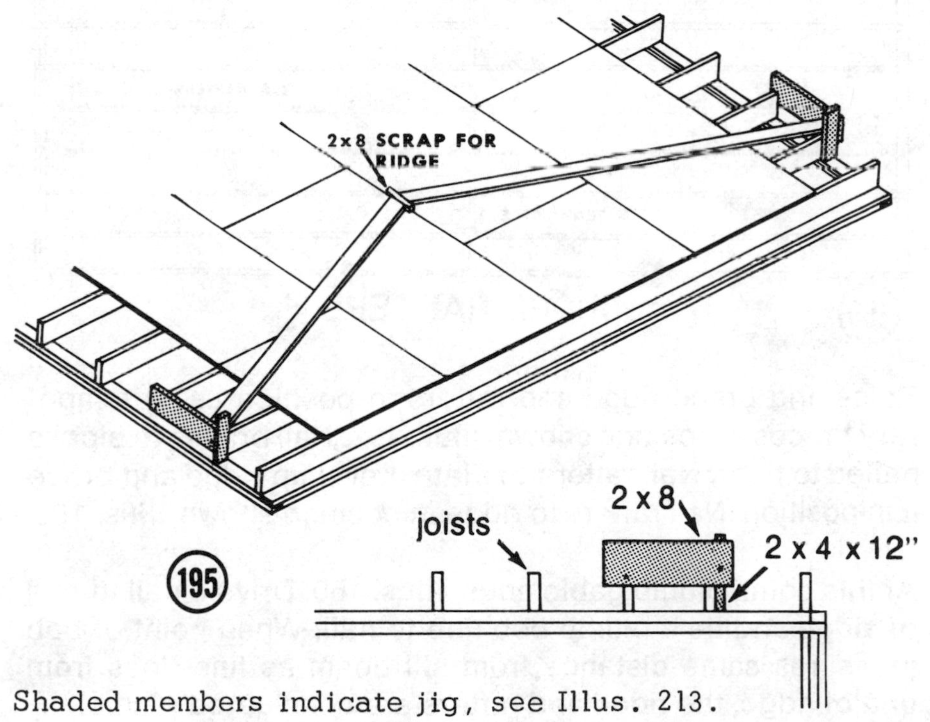

Shaded members indicate jig, see Illus. 213.

Since the rise per foot on many other roof pitches frequently runs in fractions of an inch, cutting a test pair long permits recutting to exact length required.

Mark location of rafters on both sides of one length of ridge and on both plates. Nail ridge to two No. 1 rafters, Illus. 196.

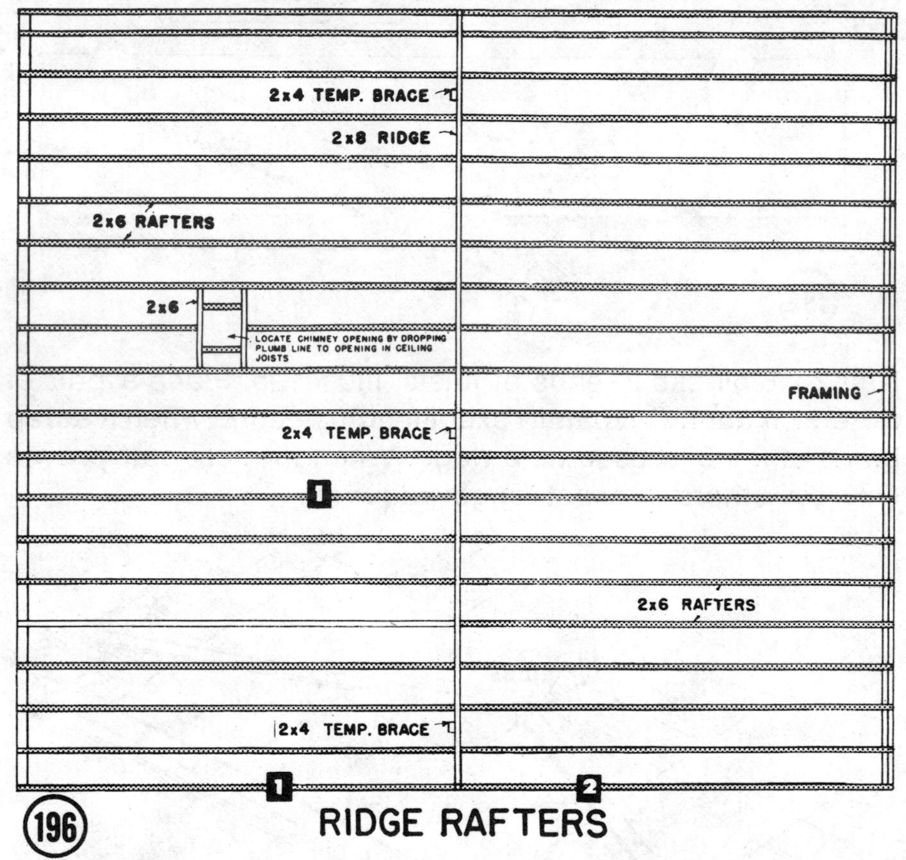

196 RIDGE RAFTERS

Raise and brace ridge and rafters in position using temporary braces in position shown, Illus. 197. Nail braces to blocks nailed to floor. Nail rafters to plate. Level up ridge and brace it in position. Nail rafters to ridge in position shown, Illus. 196.

At this point, plumb gable ends, Illus. 169. Drive a nail in end of ridge and tie a plumb bob line to nail. When point of bob measures same distance from sill beam as line does from end of ridge, the ridge and rafters are considered plumb.

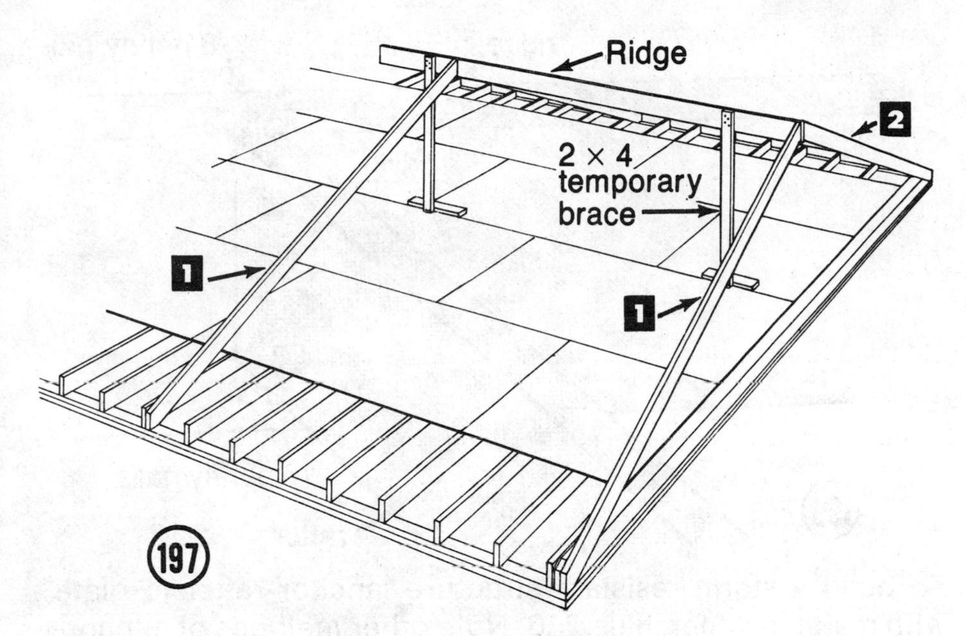

Brace ridge and rafters in plumb position, then install additional length of ridge. Nail a gusset plate, Illus. 198, to both sides of ridge where it butts together.

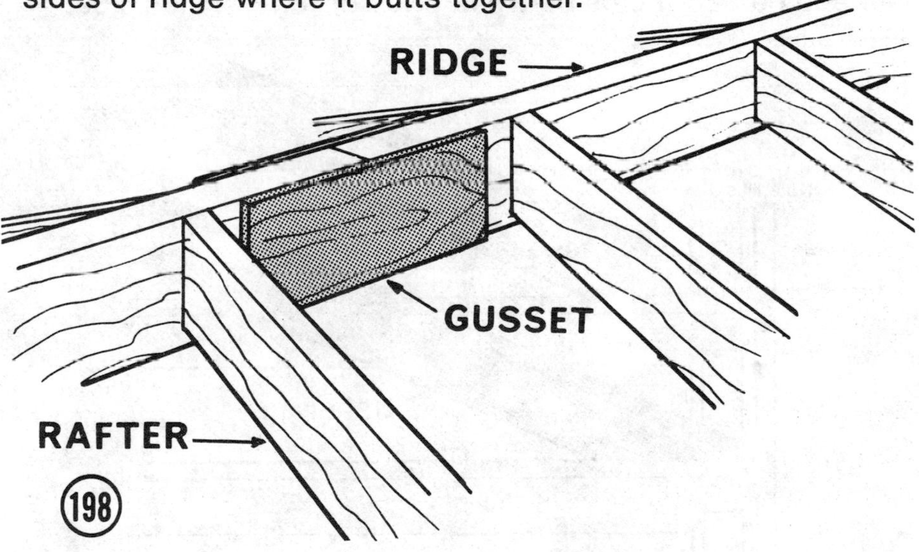

Toenail ridge to rafter with one nail, Illus. 199, then spike rafter to ridge with at least two 16 penny nails on each side. Nail rafter to plate. Nail rafter to ceiling joist, Illus. 178.

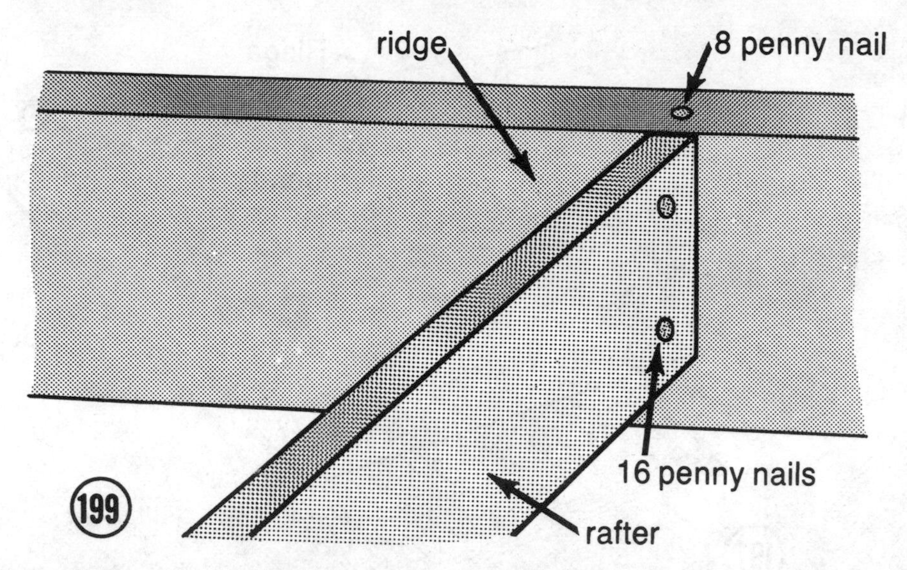

To build a storm resistant structure, anchor rafter to plate with metal fastener, Illus. 236. Note other methods of anchoring rafters on page 162.

Cut and nail 2 x 4 cats between ends of rafters, Illus. 200.

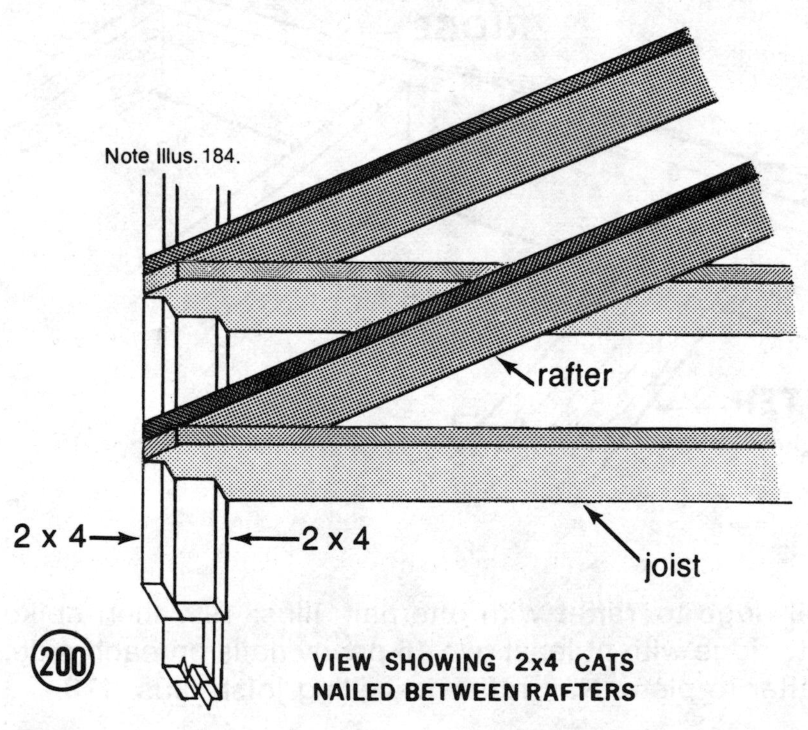

VIEW SHOWING 2x4 CATS NAILED BETWEEN RAFTERS

FRAME FOR LOUVER

Frame opening for an adjustable louver, Illus. 201, to rough opening size dealer suggests for louver selected. Cut ends of header to angle rafter requires. Check with level and toenail to rafter. Louver is installed after subsheathing has been applied.

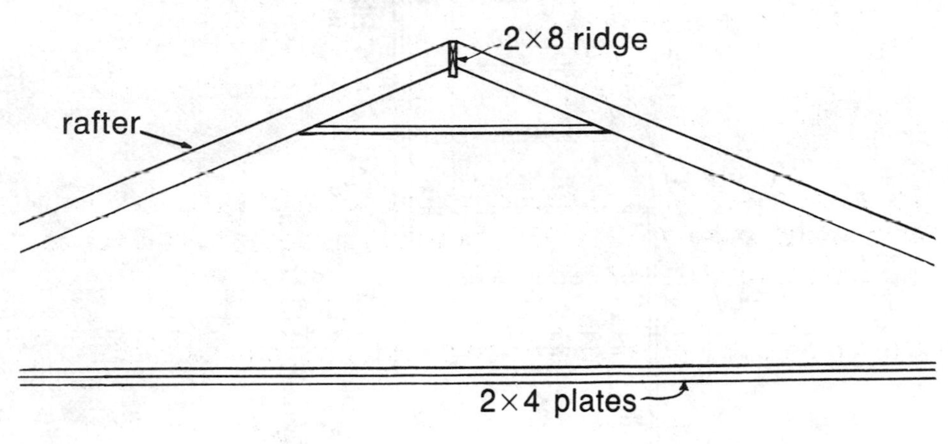

GABLE STUDS

The next step is to nail studs in gable end, Illus. 202. Cut top end to angle rafter requires and to length needed to be nailed in position shown. Nail to plate and rafter 16″ on centers with 8 penny common nails.

If framing plans specify rafter studs, Illus. 203, these are installed to stiffen rafters. First nail bridging in position required using same size lumber as ceiling joists. After nailing joists to bridging lay plyscord flooring. Nail 2 x 4 shoe in position rafter studs require. Toenail 2 x 4 rafter studs to shoe, using 8 penny nails. Spike the stud to side of rafter with 16 penny nails.

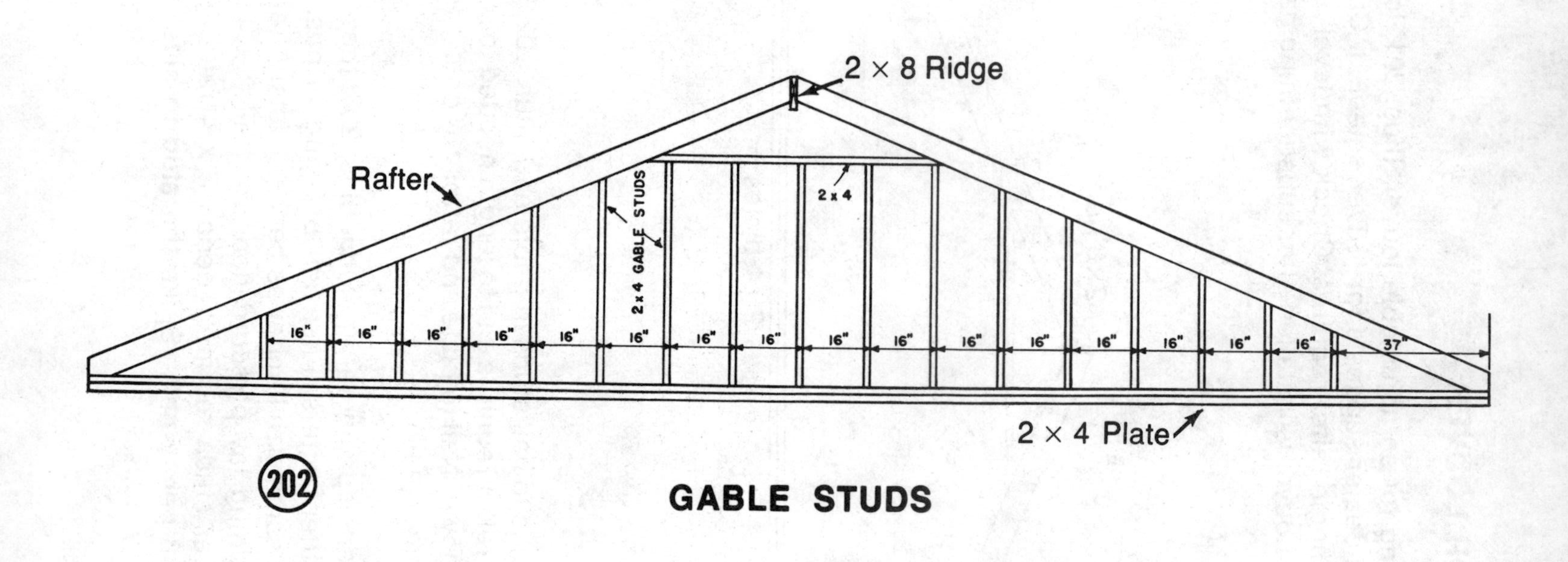

202

GABLE STUDS

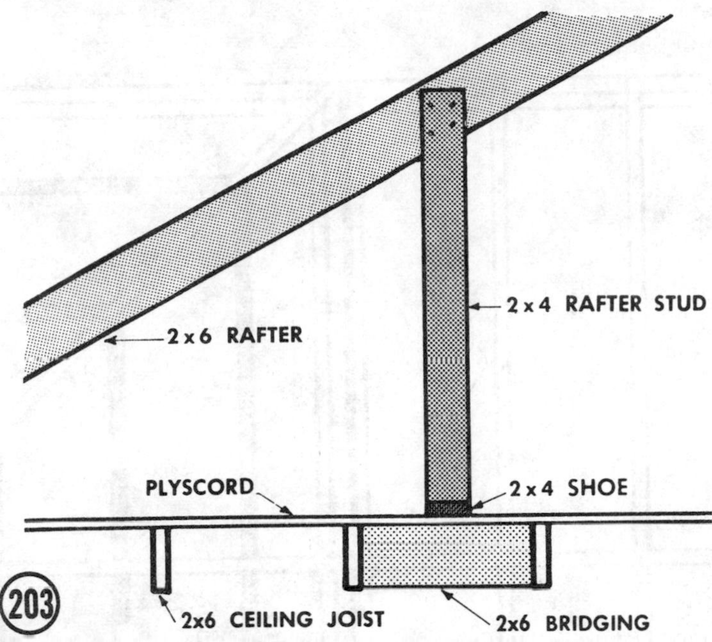

203

In this L-shaped one story ranch, Illus. 204, framing for a valley is shown in Illus. 205.

204

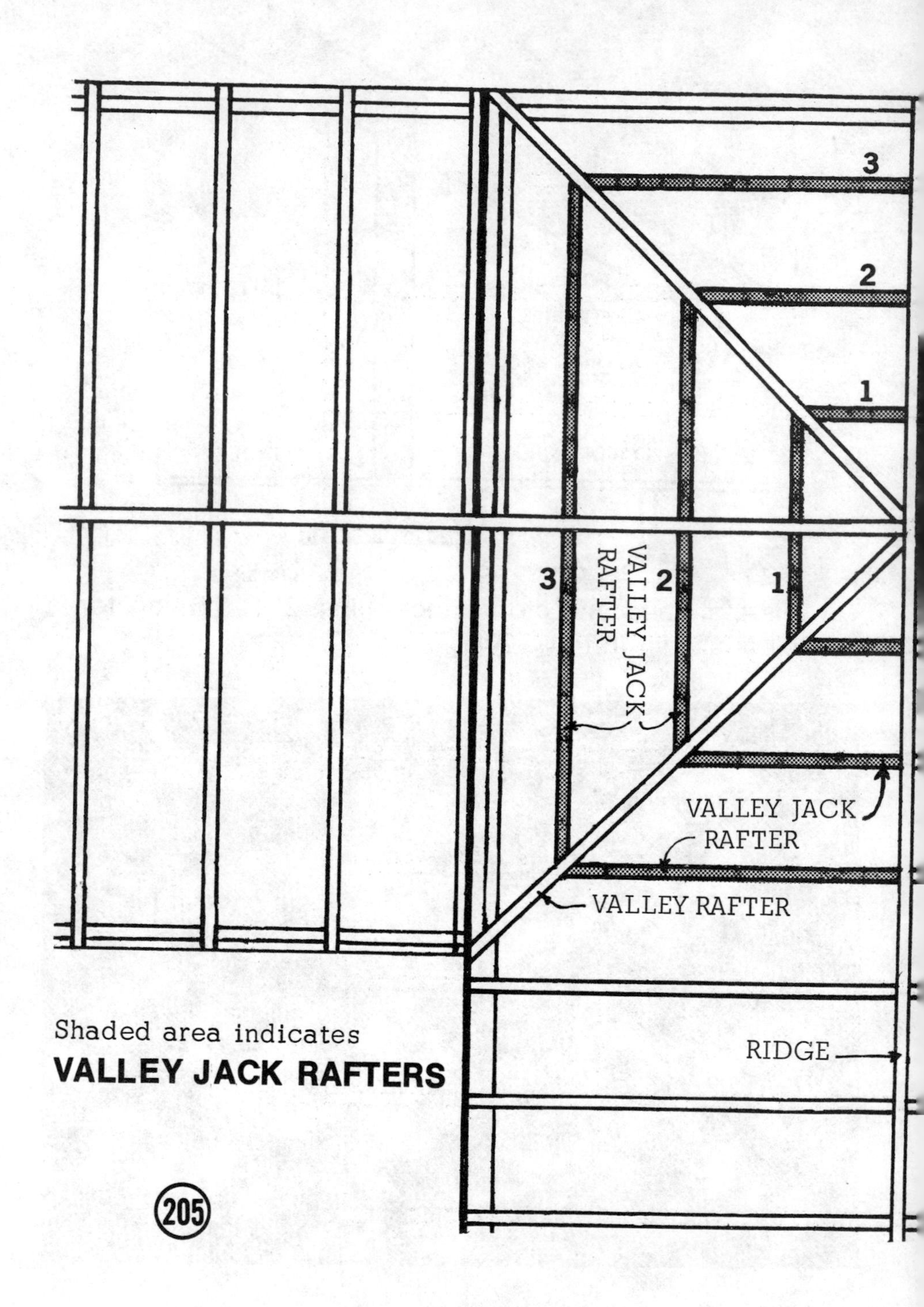

Shaded area indicates
VALLEY JACK RAFTERS

Valley jack rafters Nos. 1, 2, 3, were cut to angle shown, Illus. 206.

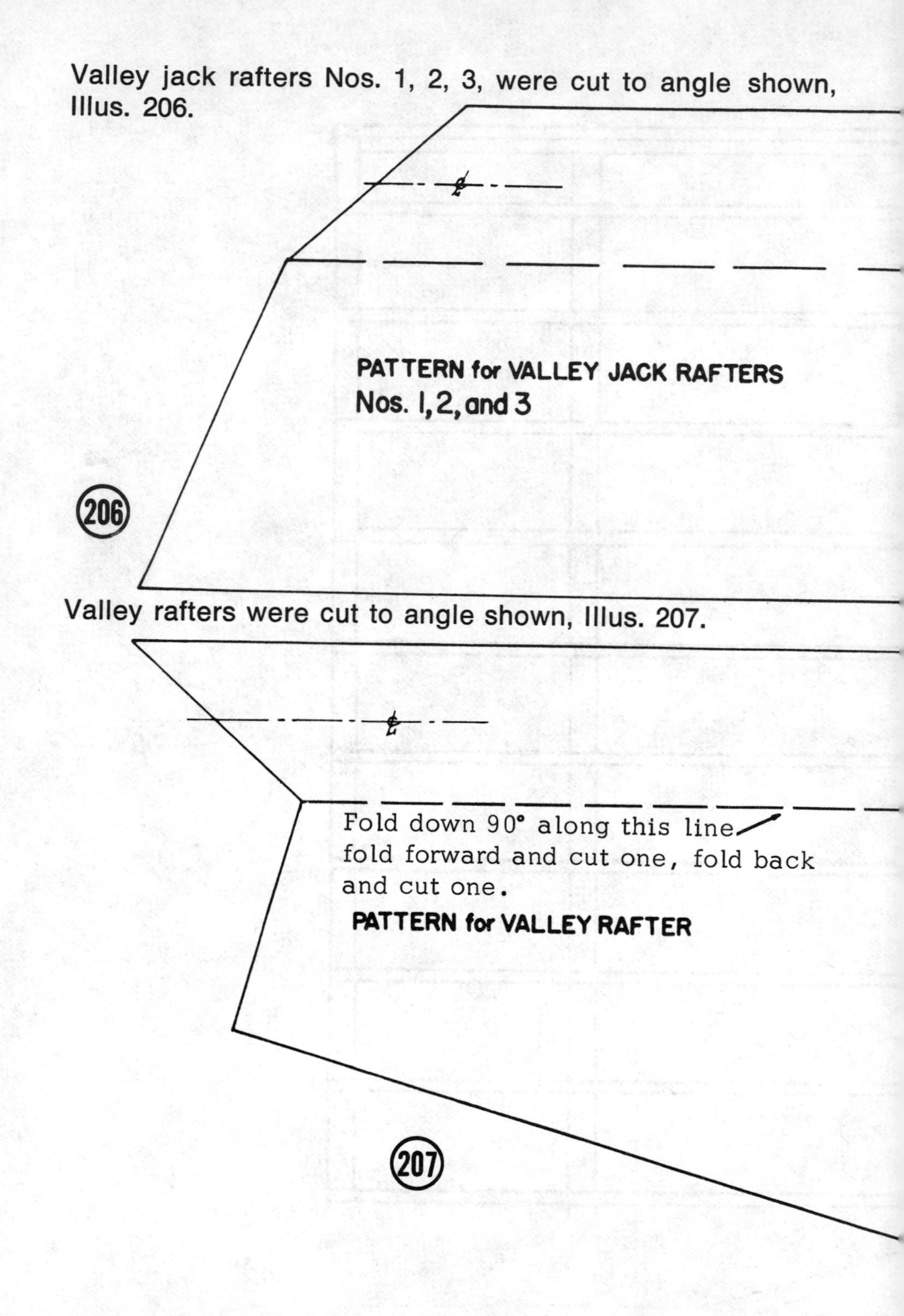

Valley rafters were cut to angle shown, Illus. 207.

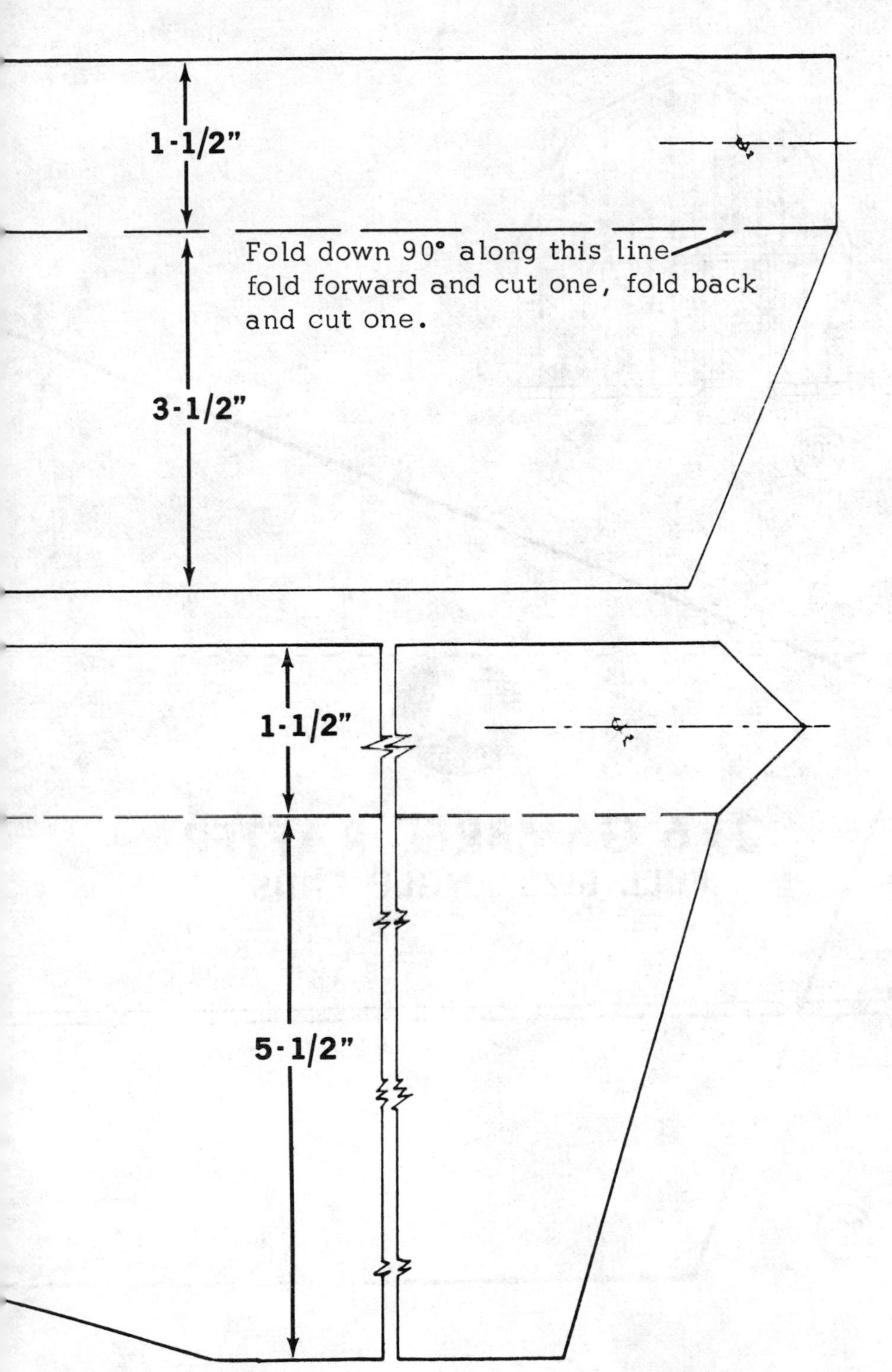
1-1/2"
Fold down 90° along this line
fold forward and cut one, fold back
and cut one.
3-1/2"
1-1/2"
5-1/2"

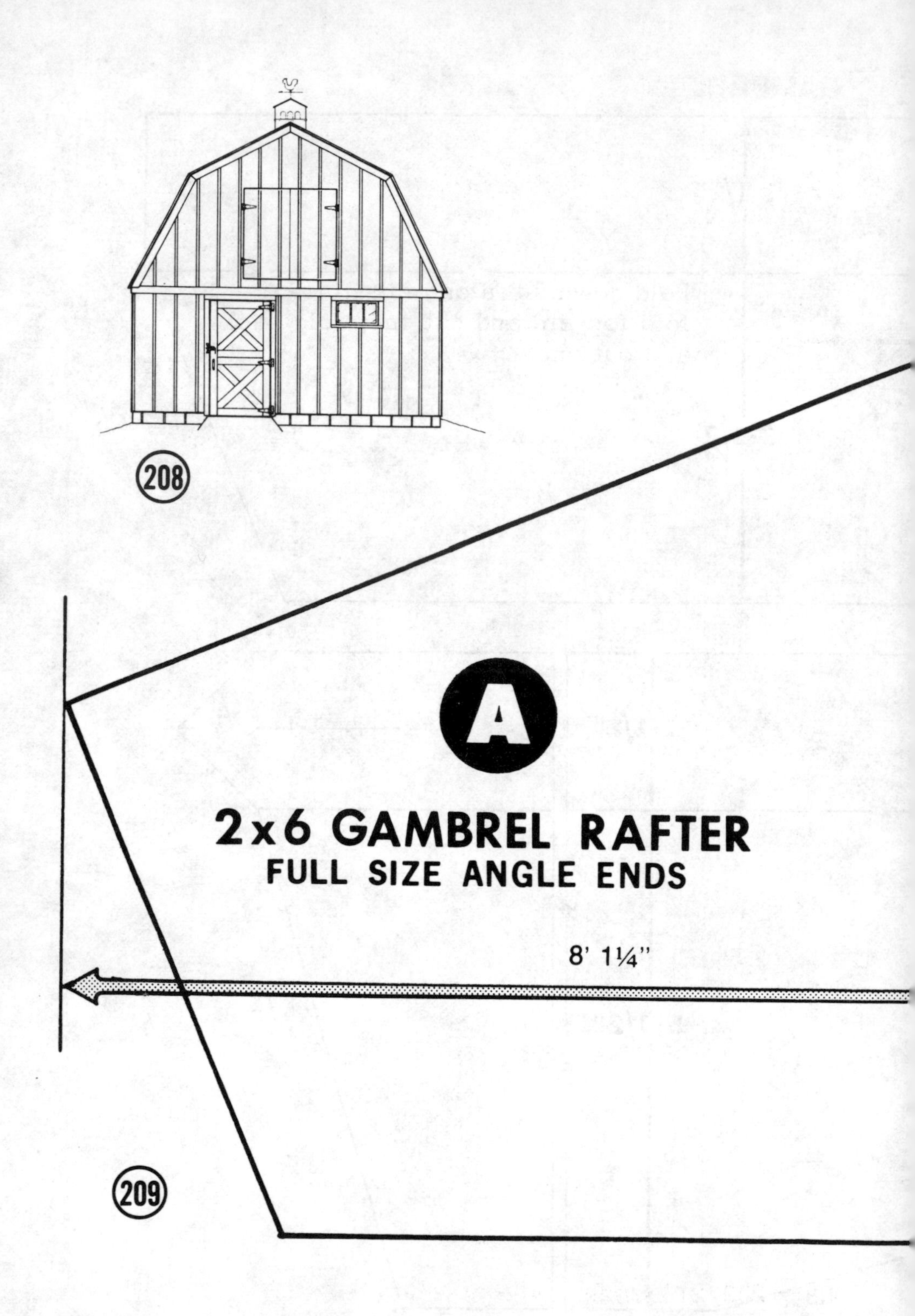
208
A
2x6 GAMBREL RAFTER
FULL SIZE ANGLE ENDS
8' 1¼"
209

GAMBREL ROOF RAFTER

If you want to build a gambrel roof rafter, Illus. 208, leave 2′0″ at the end of joists exposed, Illus. 195. Nail a 2 x 4 x 12″ vertically in position at one end. Nail a short 2 x 8 horizontally to ends of joists. Always use 1 x 6 to make a test pair of rafters. Try the test pair in place with a scrap of 2 x 8 for ridge.

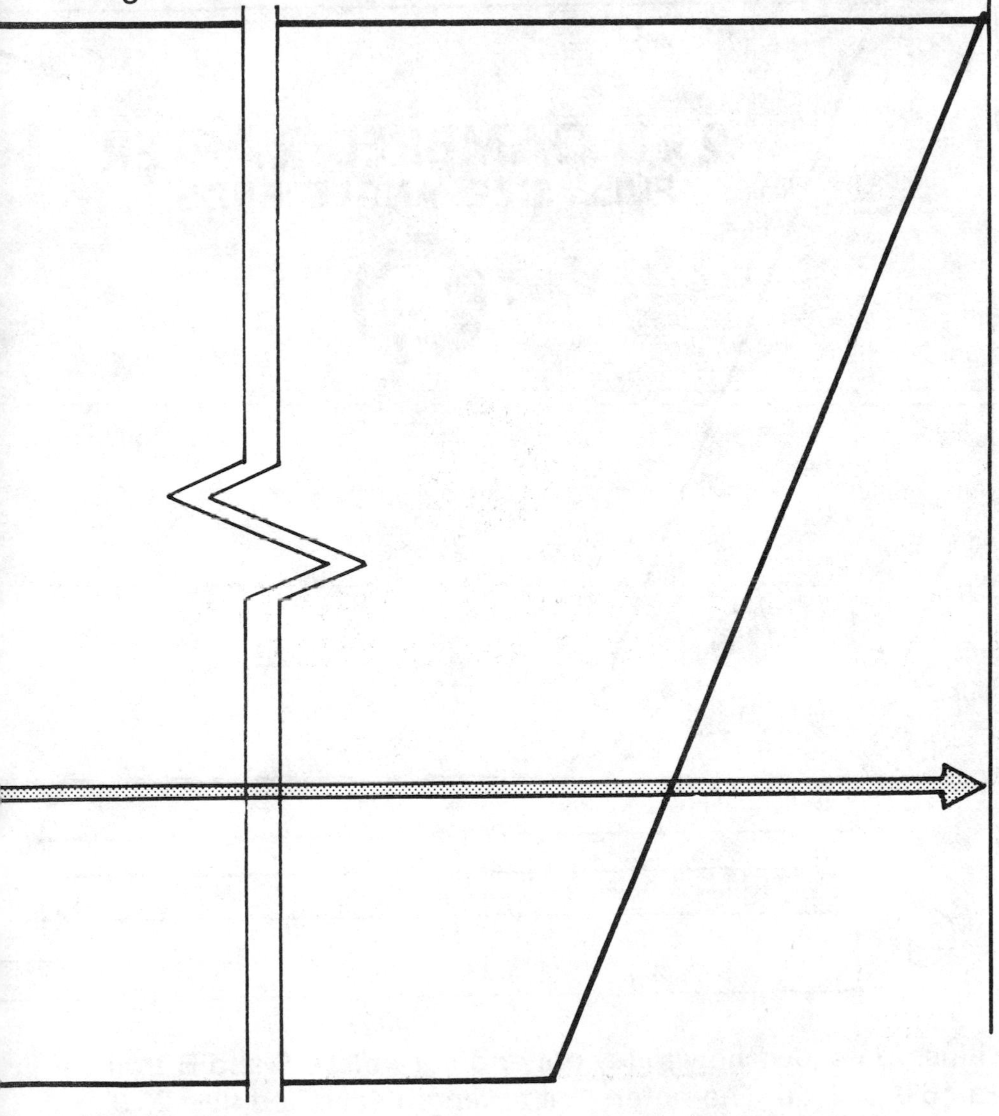

Illus. 209, 210, show angle of rafter A and B.

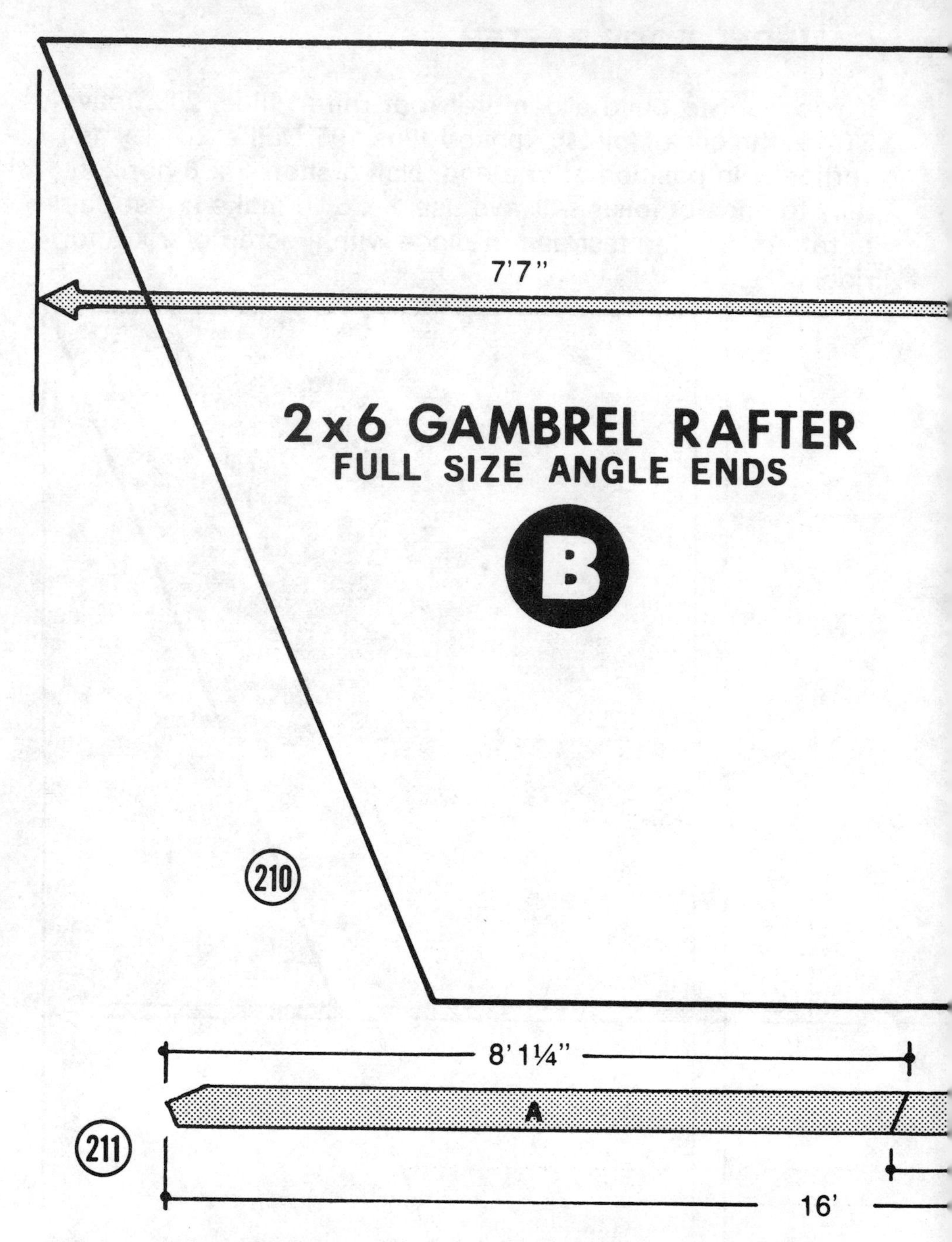

Illus. 211 shows how to lay out and cut rafters A and B from a 16'0" x 2 x 6. The rafters cut to length specified fit a 20'0" span when used with a 2 x 8 ridge.

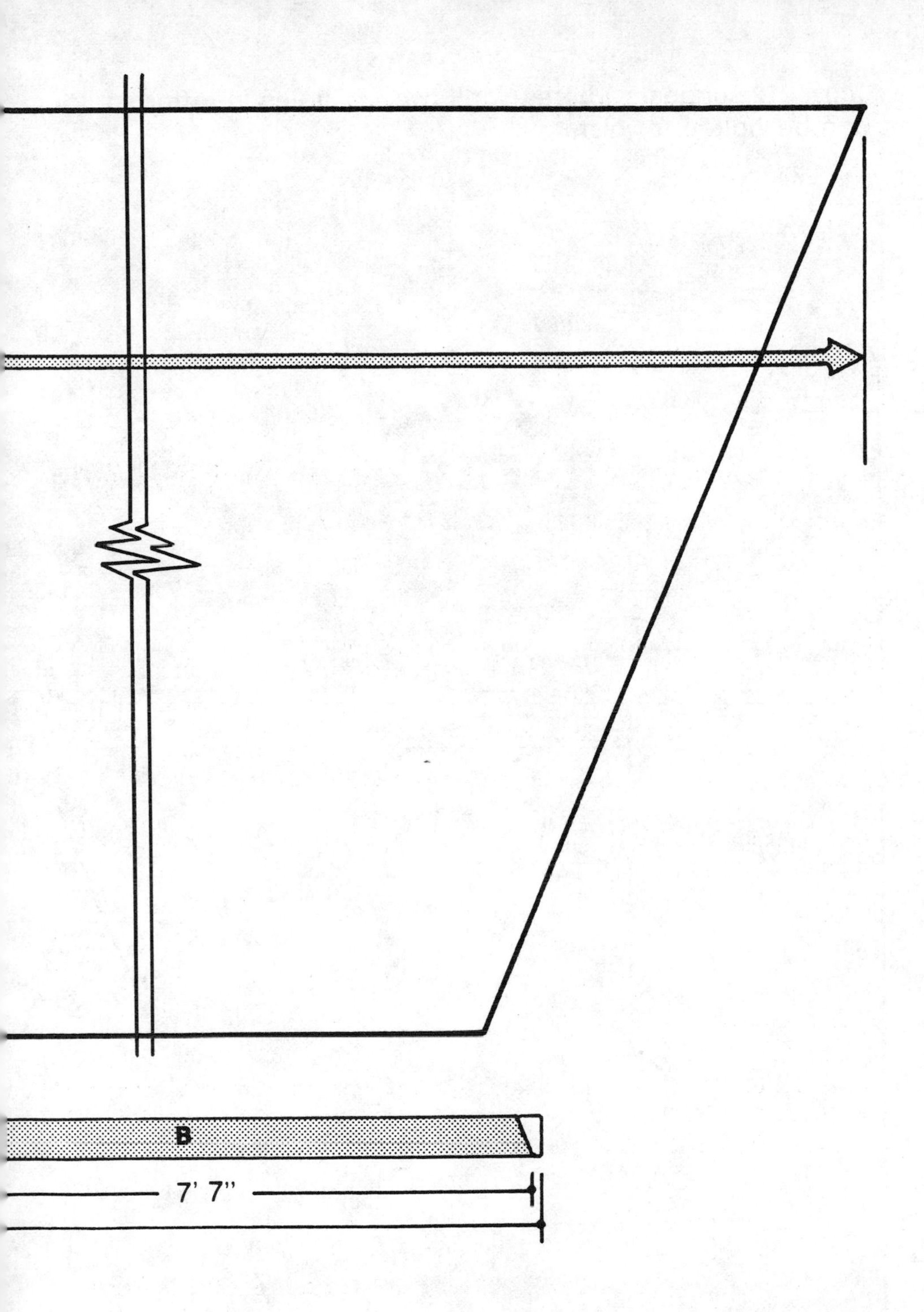
B
7' 7"

Illus. 212 suggests where to drill two ½″ holes in rafter so it can be bolted to joist.

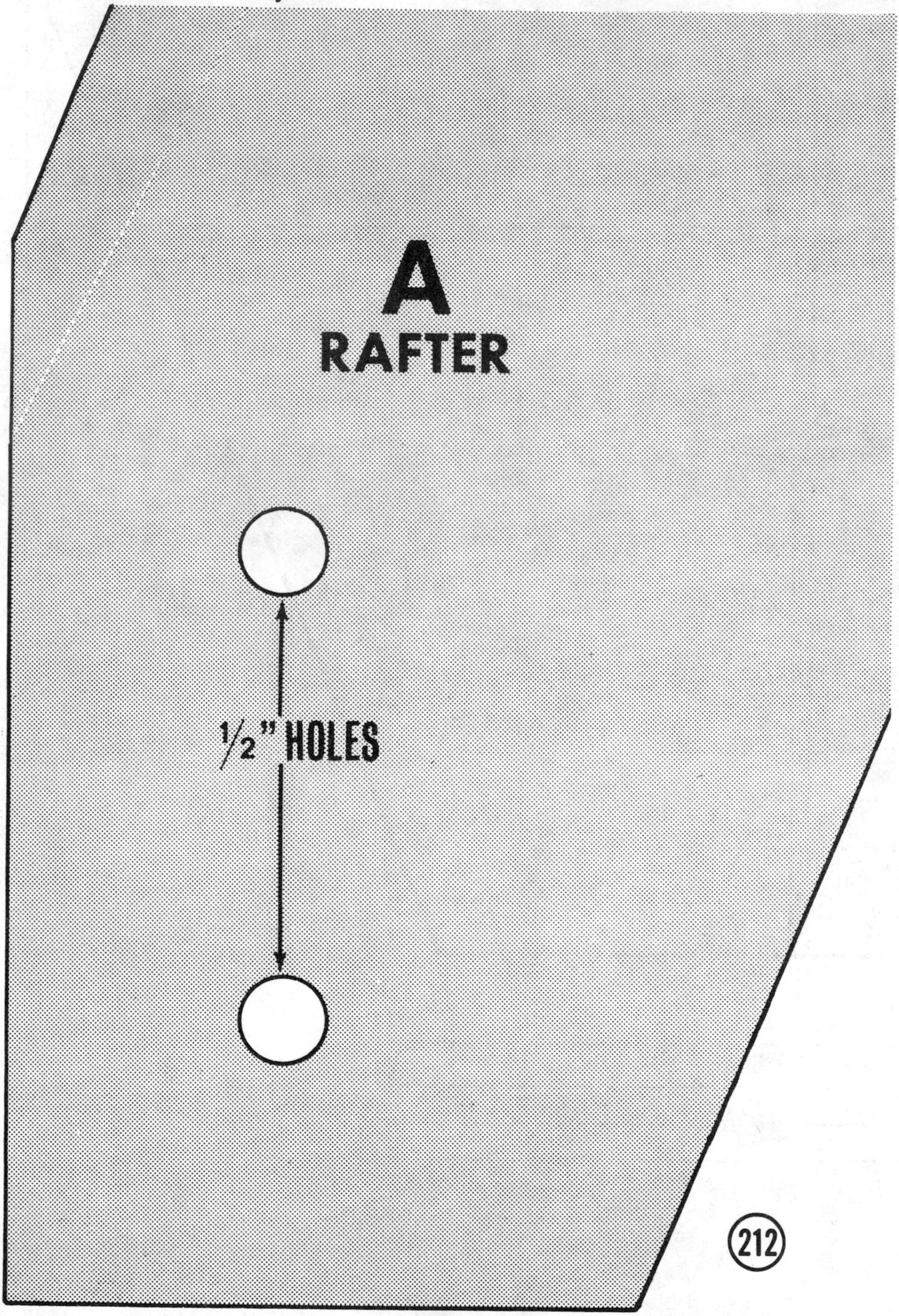

A
RAFTER

½″ HOLES

212

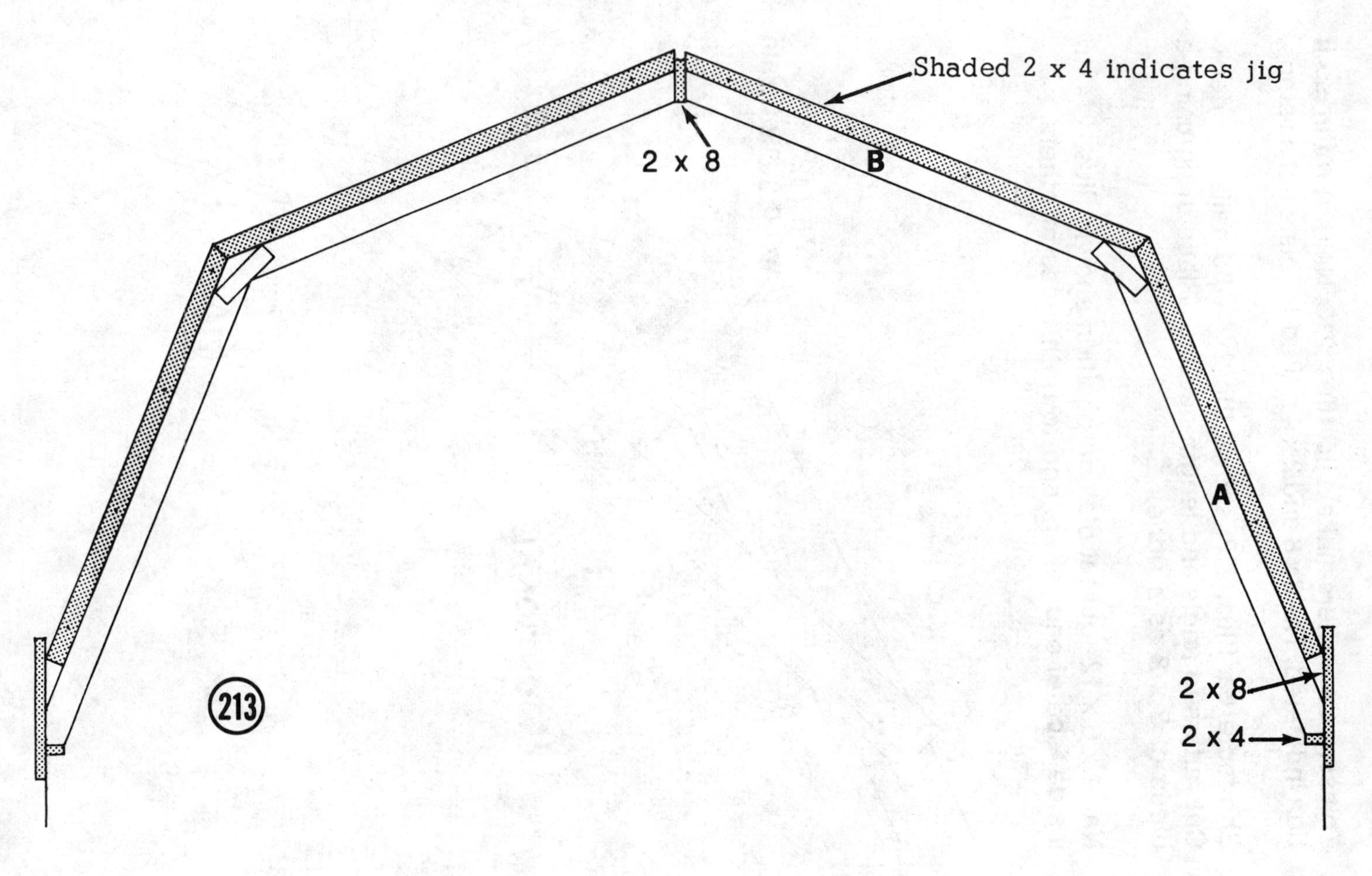
Shaded 2 x 4 indicates jig
2 x 8
B
A
2 x 8
2 x 4
213

To assemble rafters make a jig, Illus. 213. Nail 2 x 4's in position indicated. The 2 x 8 and 2 x 4 nailed to joists act as stops.

Shaded area, Illus. 213, shows the 2 x 4 jig nailed to deck. Cut rafters to angle and length needed. Place in jig using a piece of 2 x 8 as a ridge.

Nail 1 x 6 x 12″ gusset plate in position shown, Illus. 214, to inside face of end rafters and to both faces of others.

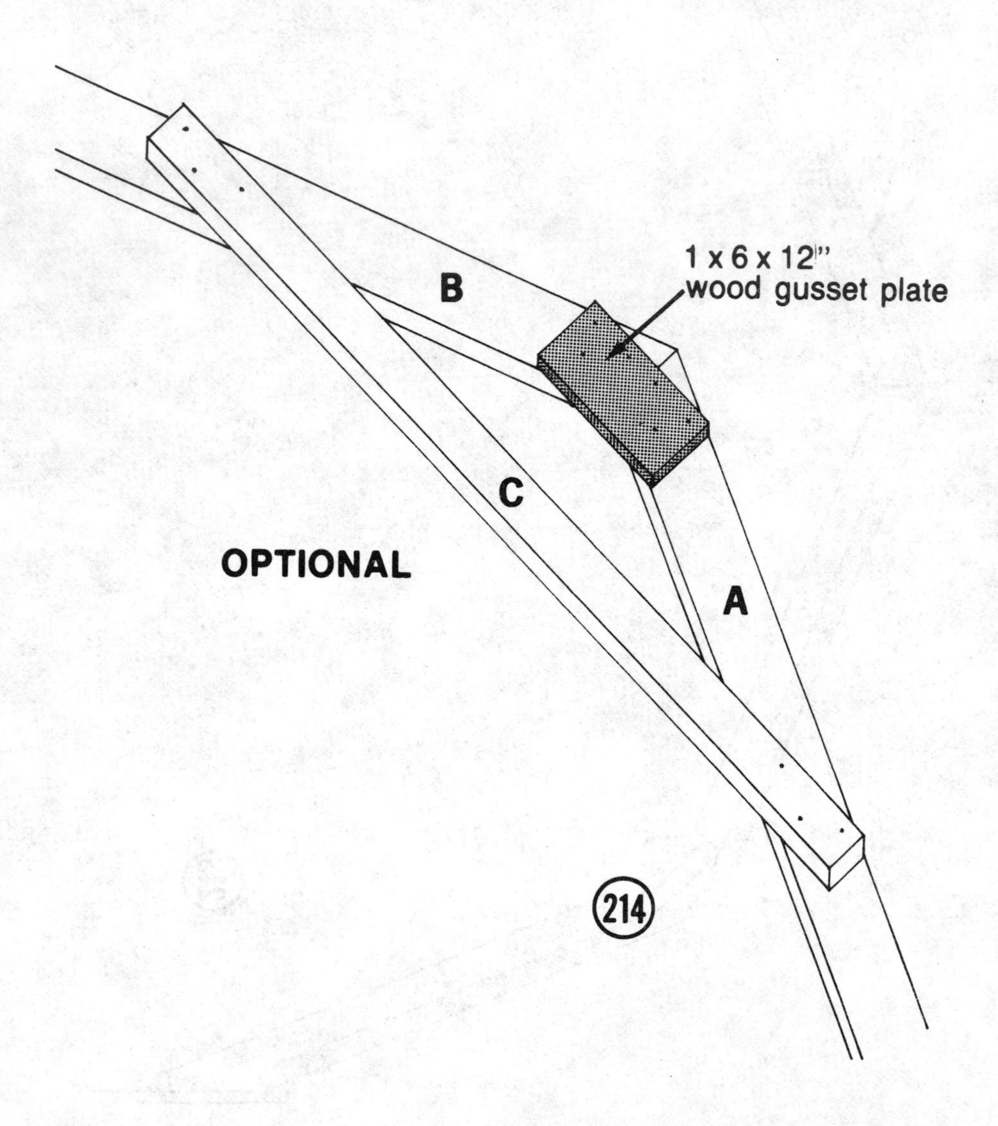

Metal plates, Illus. 215, can be used in place of wood. If metal plates are used, these can be applied to both faces of all rafters.

Nail 2 x 4 brace C, in position.

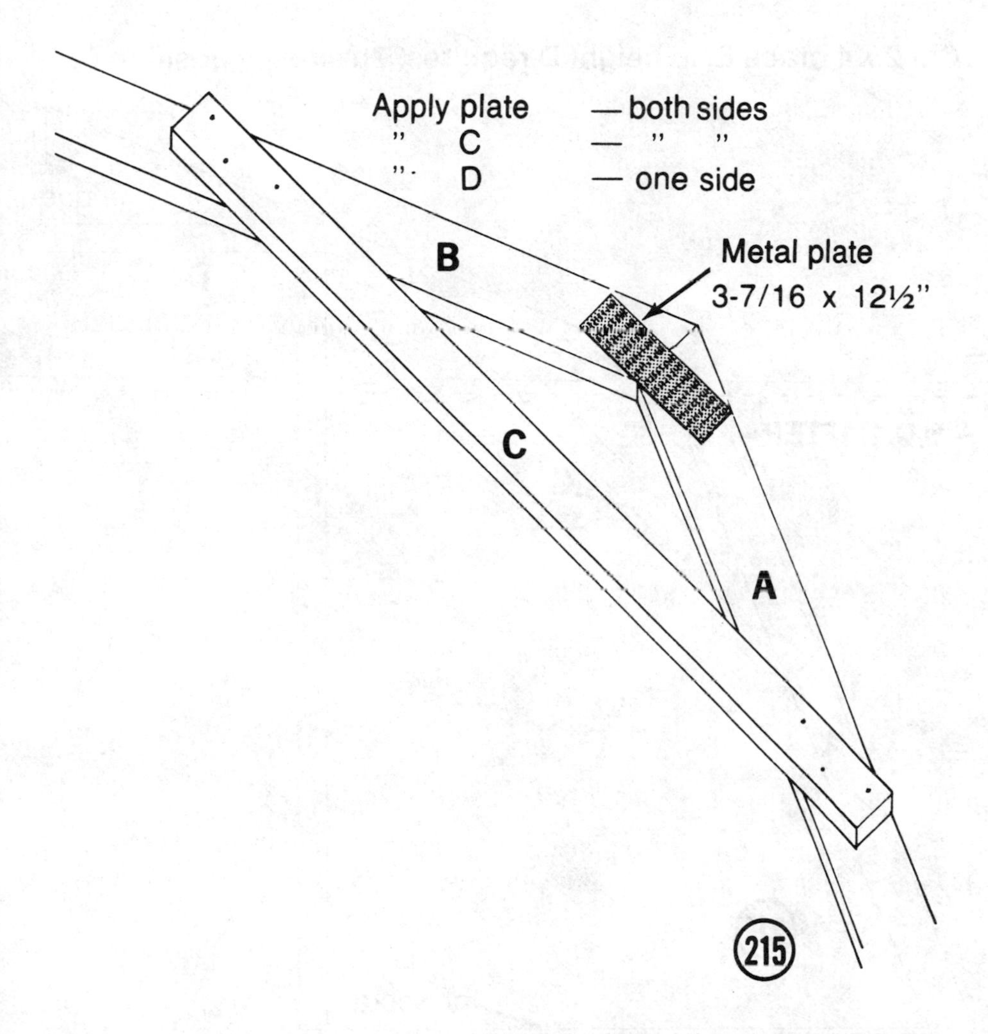

Collar beam D, Illus. 216, is nailed in position shown after all rafters have been raised and roof sheathed. Always apply collar beam on opposite side of cross brace C.

Place rafter A in position. Using holes in A as a guide, drill ½″ holes through joist. After assembling A, B and C, A can be bolted to joist and braced temporarily with E until nailed to ridge.

Cut 2 x 4 brace E, to height D requires. This helps position D.

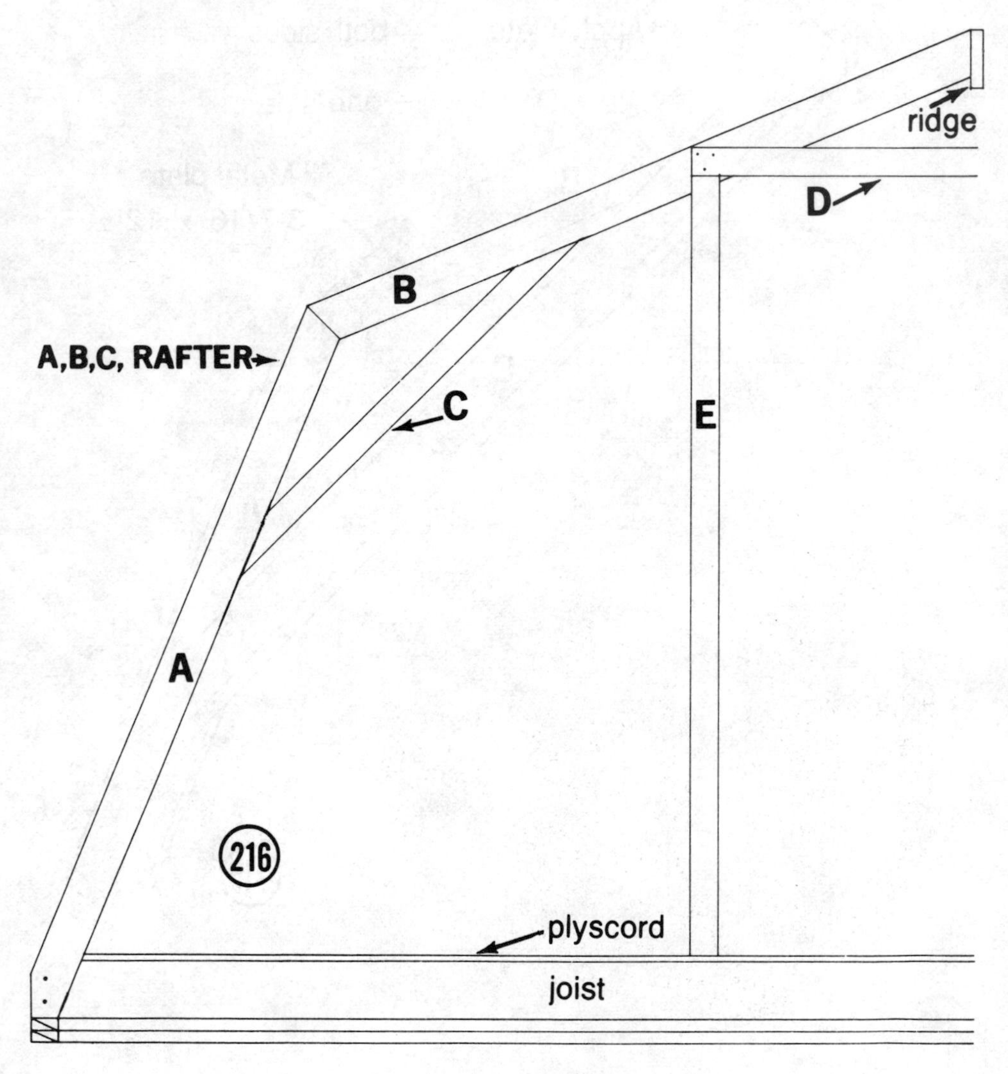

TRUSS RAFTERS

Most lumber retailers now sell prefabricated truss rafters, Illus. 217. Many, using 2 x 6, are engineered to meet all codes. The truss shown is approved for 28-30 ft. spans — 5 in 12 pitch. Prefabricated metal truss fasteners, Illus. 217, are used to assemble truss. Special hangers, Illus. 218, are available when fastening joist to a header.

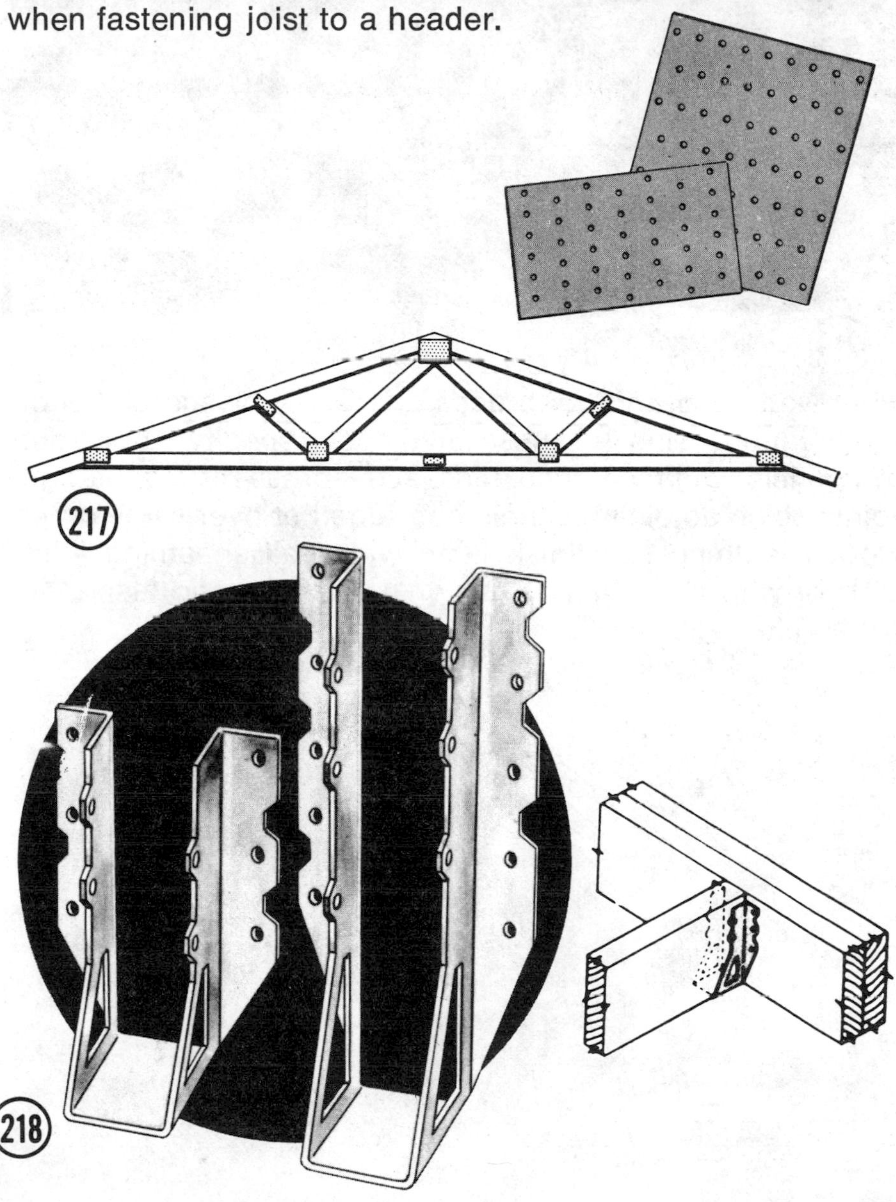

1 x 4 furring, Illus. 219, nailed 16″ on centers to bottom of truss, permits nailing ½″ gypsum ceiling panels.

219

Note Illus. 151a.

Remove temporary roof braces and apply plyscord sheathing to rafters using thickness local codes specify. Apply roofing panels with face grain, Illus. 220, across rafters. Stagger joints so no adjoining panels butt together over same rafter. Roof sheathing can finish flush with wall sheathing, Illus. 221; or with fascia, Illus. 222; or project beyond fascia ¾″ to 1″, Illus. 223.

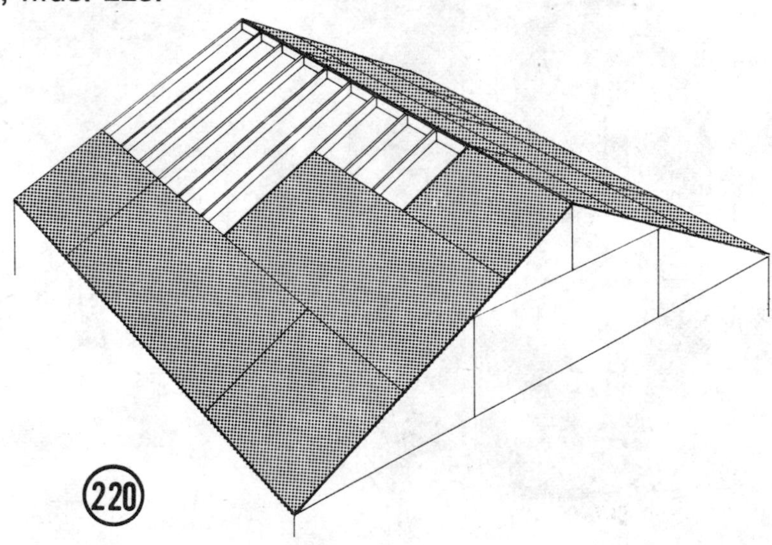

220

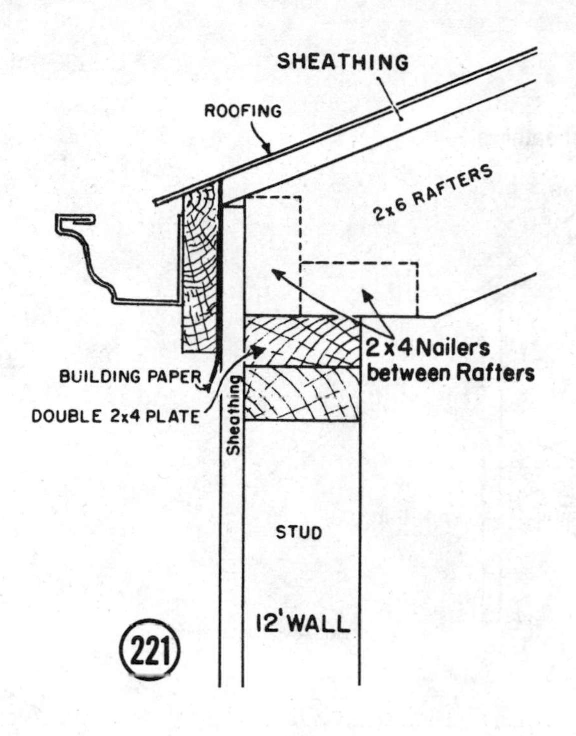
SHEATHING
ROOFING
2x6 RAFTERS
2x4 Nailers
between Rafters
BUILDING PAPER
DOUBLE 2x4 PLATE
Sheathing
STUD
12' WALL
221

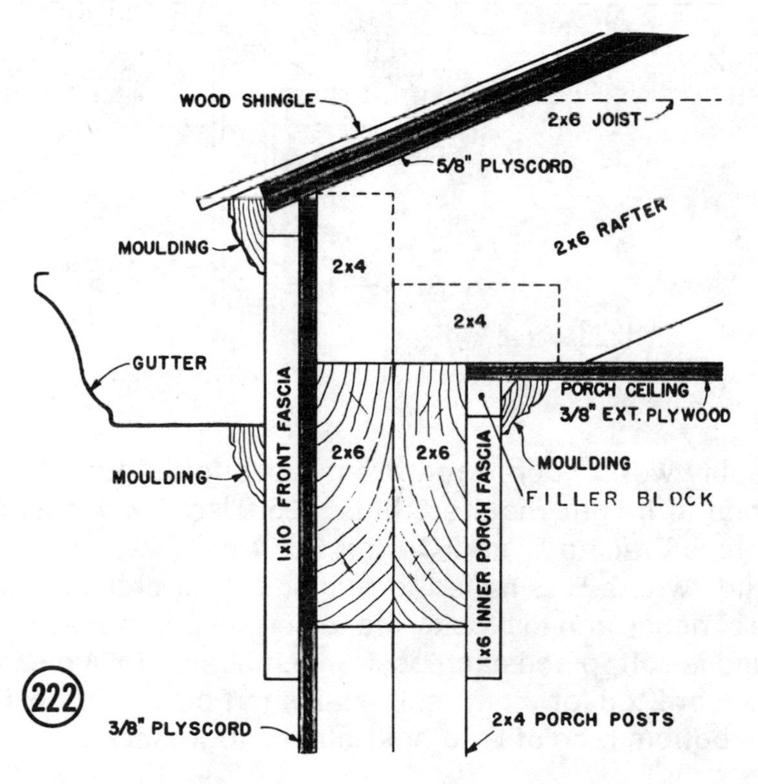
WOOD SHINGLE
2x6 JOIST
5/8" PLYSCORD
MOULDING
2x4
2x6 RAFTER
2x4
GUTTER
PORCH CEILING
3/8" EXT. PLYWOOD
1x10 FRONT FASCIA
2x6
2x6
MOULDING
MOULDING
FILLER BLOCK
1x6 INNER PORCH FASCIA
222
3/8" PLYSCORD
2x4 PORCH POSTS

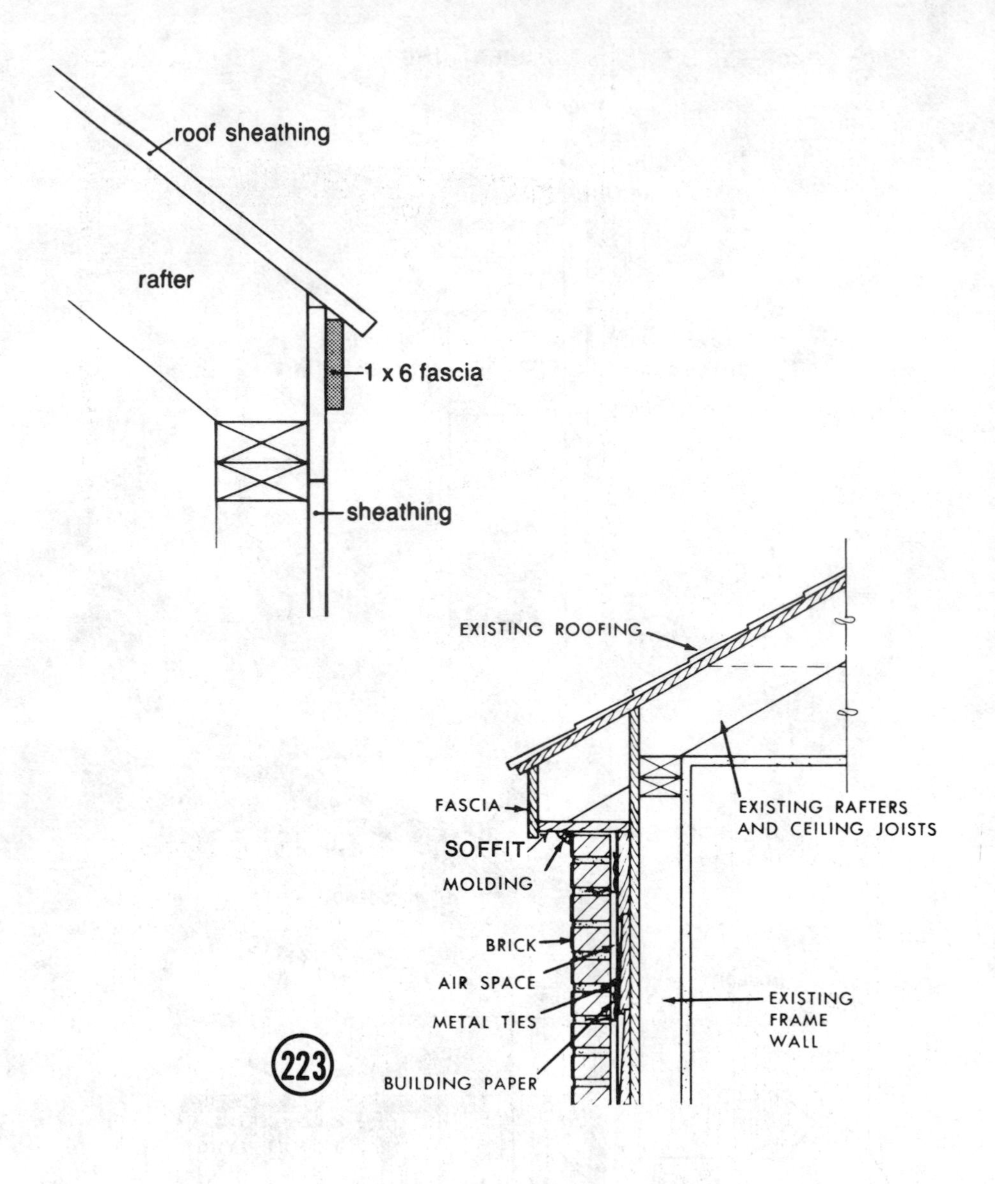

To simplify working on a roof, rent a scaffold, Illus. 224, or build one to height required, Illus. 225. Use 2 x 4 posts A, 1 x 6 cross bracing B and C. Nail 2 x 4 blocks temporarily into studs where B is nailed to building. Nail cross-bracing to side of house and to blocks. Make certain posts are plumb. If ground is soft, place each post on a piece of 2 x 4 or 2 x 6. Use 2 x 6 or 2 x 8 for platform D. Nail short pieces of 1 x 6 or 2 x 4 to bottom face of D to hold planks together.

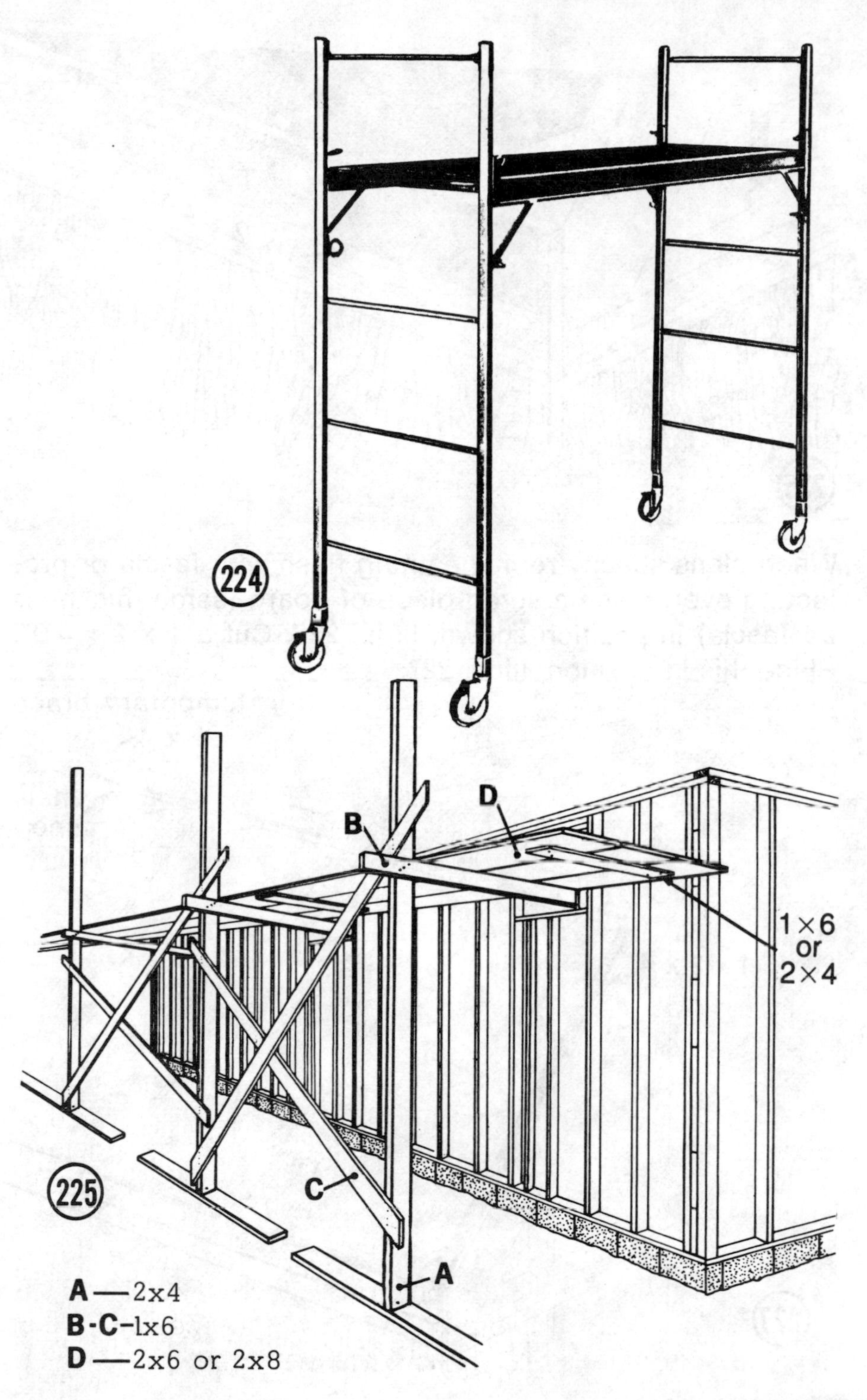
224
225
B
D
1×6
or
2×4
C
A
A —2x4
B-C–1x6
D —2x6 or 2x8

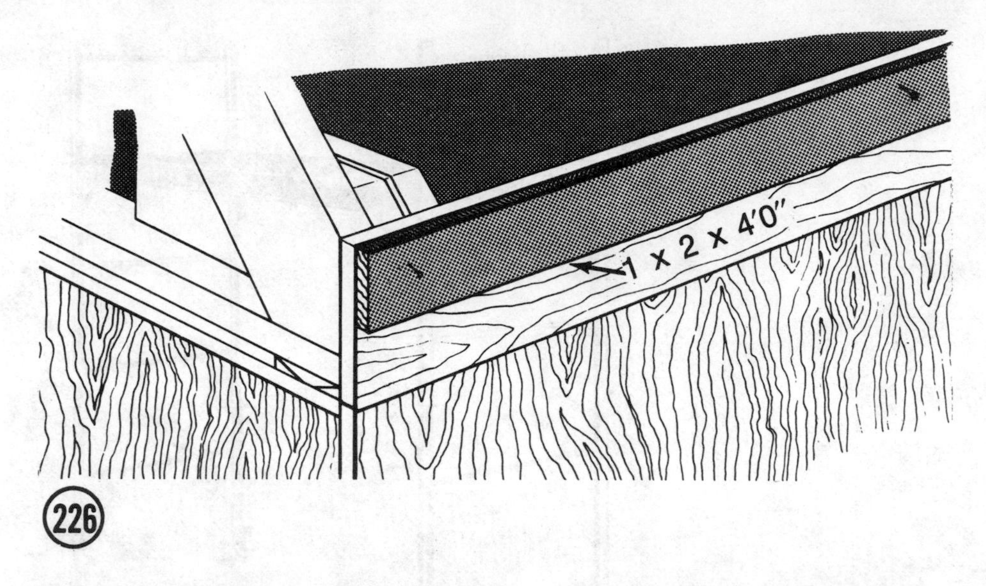

226

When plans specify roof sheathing flush with fascia or projecting over, place a scrap piece of board (same thickness as fascia) in position shown, Illus. 226. Cut a 1 x 2 x 4'0". Place this in position, Illus. 227.

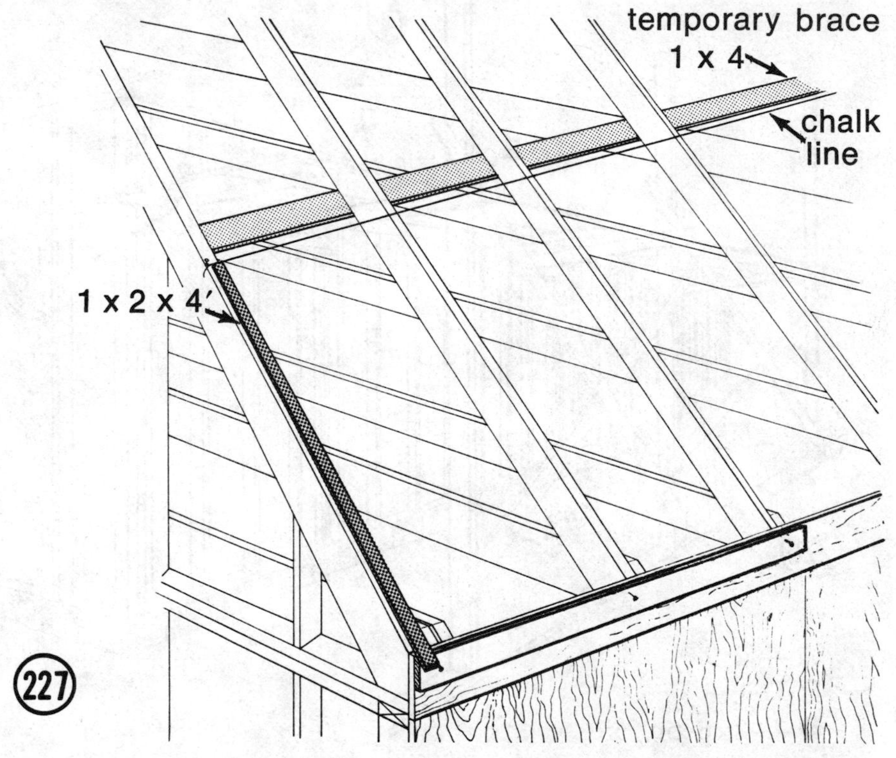

227

Allow 1 x 2 to project over face of fascia amount plans specify for roof sheathing. Draw a line on rafter. Do the same at other end of building. Snap a chalk line across rafters. Apply the first course of plyscord sheathing to this line. If you want roof sheathing to finish flush with face of fascia, position 1 x 2 accordingly. Nail sheathing to roof with 8 penny common nails.

Nail the first course of roofing panels in position plans specify working off the scaffold. After completing the first course, nail a 2 x 4 x 16′ foot support to roof, Illus. 228. This simplifies applying other panels.

To simplify raising additional panels to roof, build a frame, Illus. 229. Use two 2 x 4 x 16′ or length required for A. Nail cross braces B, a 2 x 8 C and D in position shown. Load up before going up on roof.

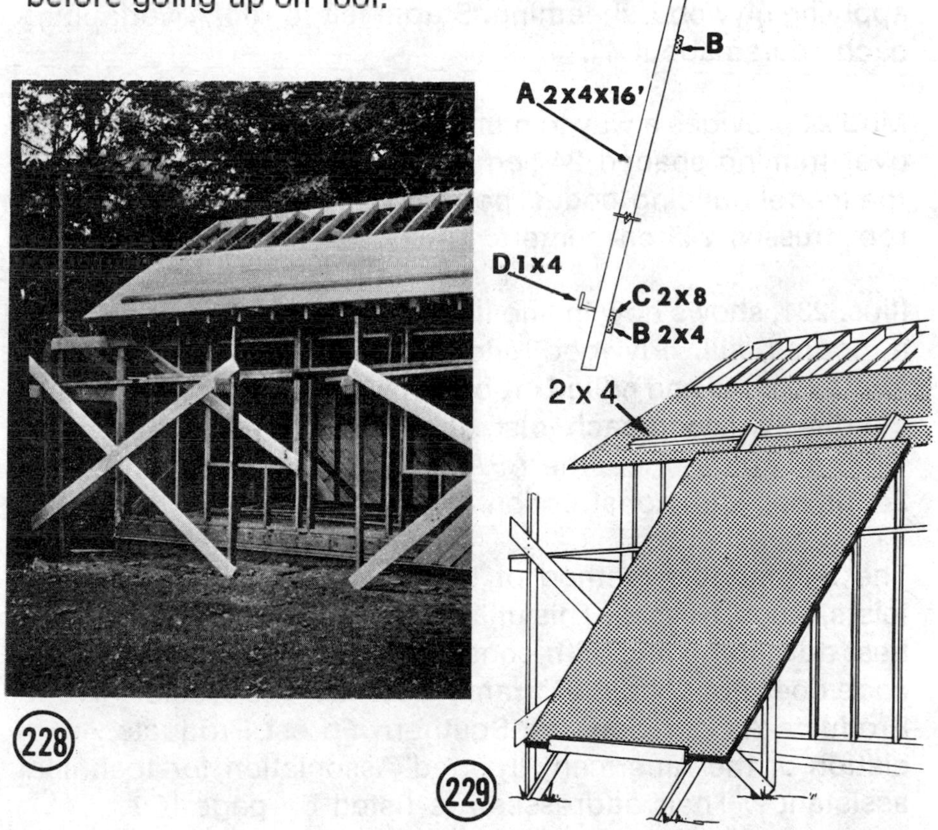

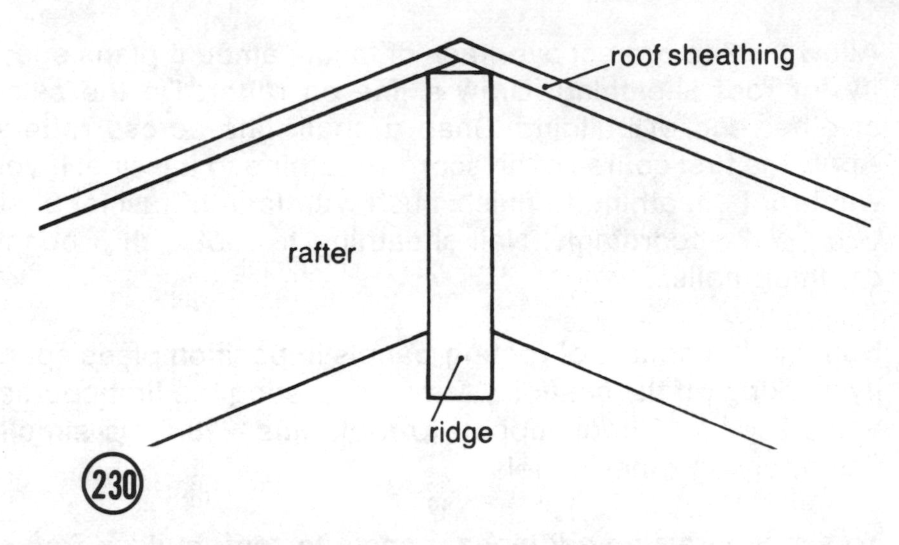

Saw panels at ridge to angle required, Illus. 230.

Cover roof sheathing with #15 felt as soon as you finish applying plywood sheathing. Staple felt to roof, overlapping each course about 4″.

Mod 24 provides a way to build better for less, using plywood over framing spaced 24″ on centers. Mod 24, approved by the model building codes, permits spacing joists, studs and roof trusses 24″ on centers.

Illus. 231, shows how in line floor joists can be installed over a poured sill. Plywood underlayment, Illus. 232, with lap joints is glued and nailed in position. Apply a continuous bead of adhesive along each joist, Illus. 141. Tested at the Underwriters Laboratories, the APA glued floor system with Mod 24 lumber joist construction has a one hour fire rating.

The 24″ module method of construction requires aligning joists, studs, trusses. This utilizes lumber and plywood to the best advantage and with considerable savings. If your local code does not accept 24″framing,contact the Western Wood Products Association, the Southern Forest Products Association or the American Plywood Association for technical assistance. Their addresses are listed on page 162.

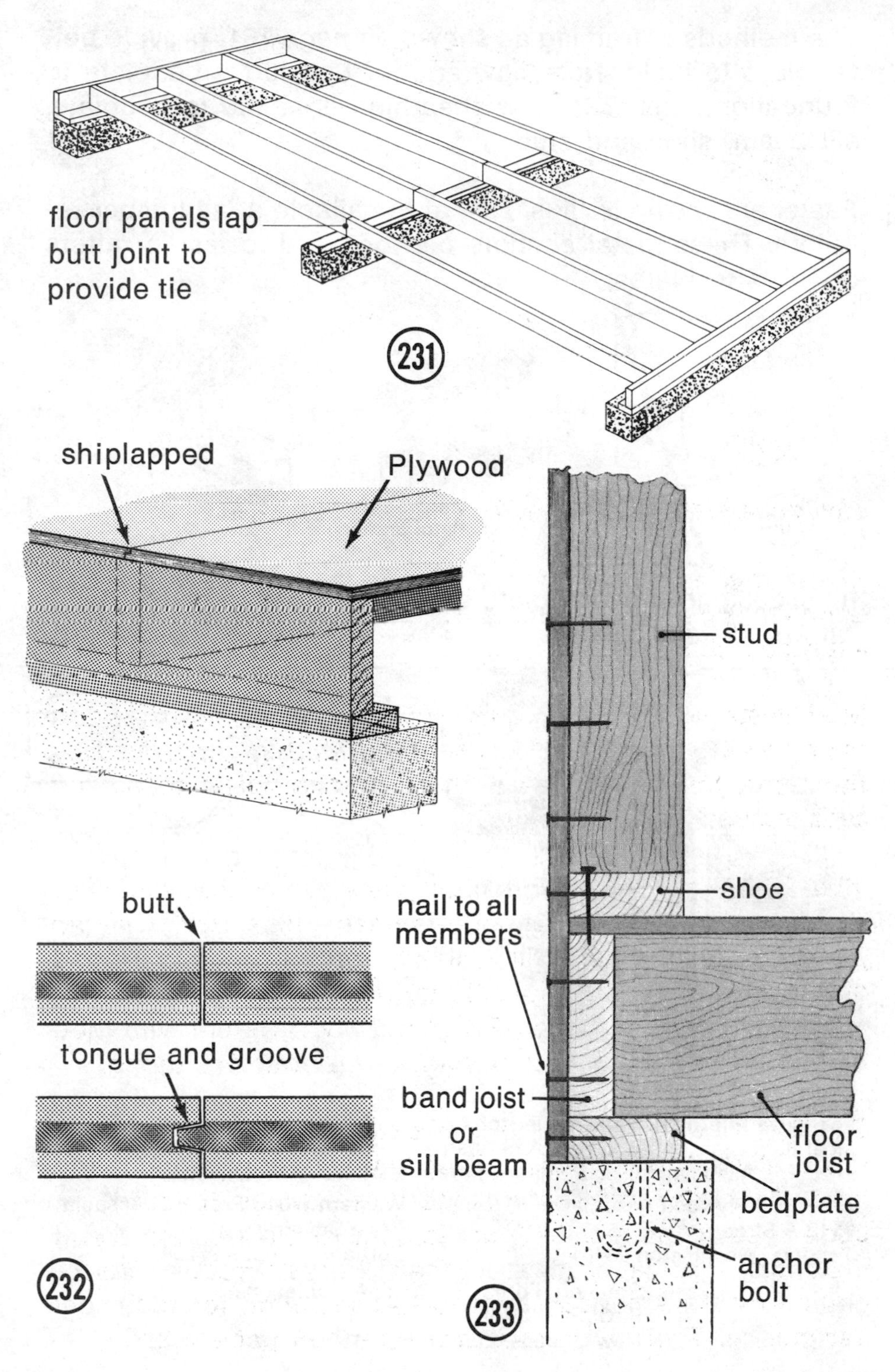
floor panels lap
butt joint to
provide tie
231
shiplapped
Plywood
stud
shoe
butt
nail to all
members
tongue and groove
band joist
or
sill beam
floor
joist
bedplate
anchor
bolt
232
233

The methods of framing as shown on page 161, provide better ways to build. Note how anchor bolt locks bedplate to foundation, Illus. 233; how sheathing is nailed to bedplate, sill beam, shoe and stud.

Fasteners shown in Illus. 234, are available at all lumber retailers. These metal anchors can be used to fasten rafters or joists to plates.

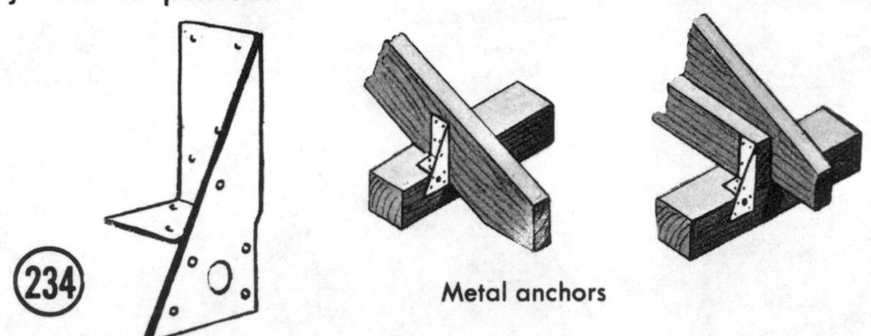

Illus. 235 shows metal plate connector used to fasten beam rafters to ridge beam.

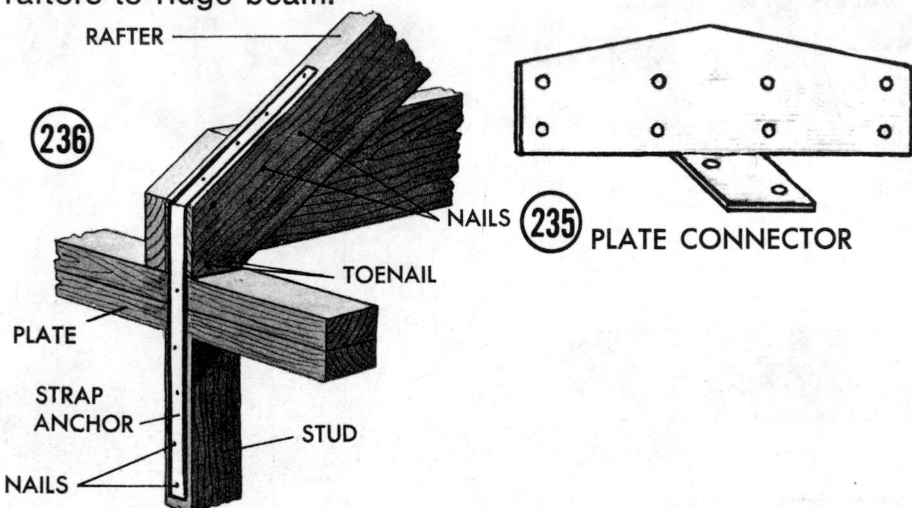

For more information on related plywood and lumber literature; or code, technical field assistance, contact American Plywood Association or Western Wood Products Association, or Southern Forest Products Association.

American Plywood Association
1119 A Street
Tacoma, Washington 98401

Western Wood Products Association
Yeon Building
Portland, Oregon 97204

Southern Forest Products Association
P. O. Box 52468
New Orleans, Louisiana 70152

HOW TO USE A LEVEL, LEVEL-TRANSIT

Every building, masonry, grading, driveway and drainage job, to mention but a few, requires establishing accurate layout lines as described on page 24. All construction requires level and plumb lines. Guide lines must be straight, level, corners square and plumb.

Just as a carpenter's square, Illus. 1, 17, helps establish a right angle, so a level, Illus. 237, a level-transit, Illus. 238, or a hand held sight level, Illus. 239, assist in sighting and establishing straight level lines and/or planes.

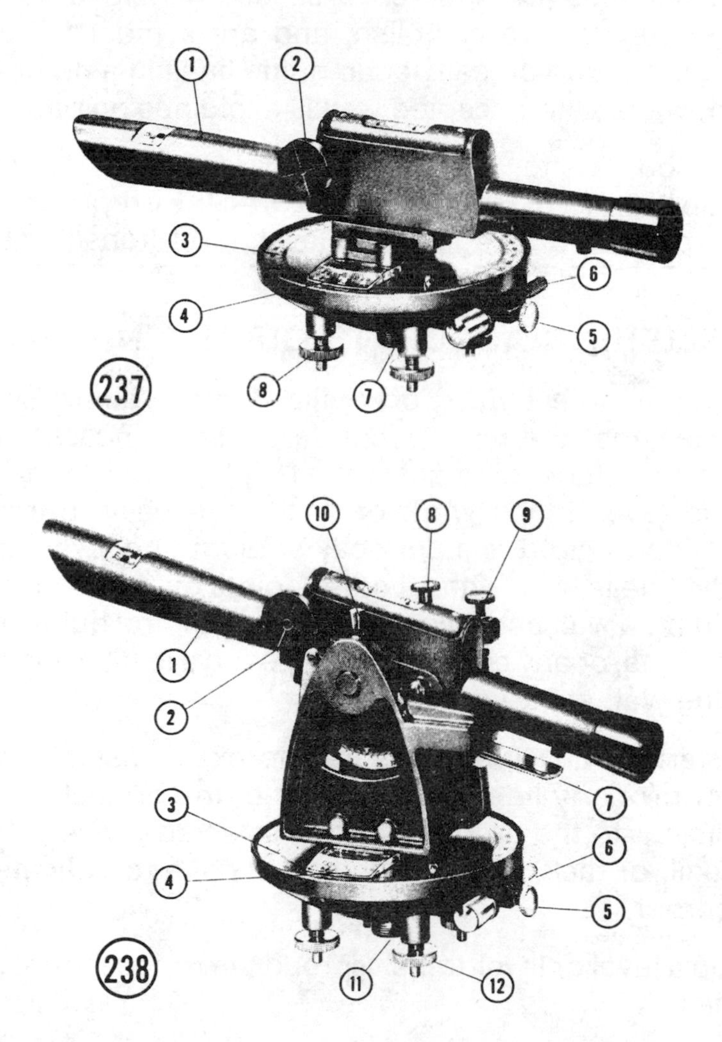

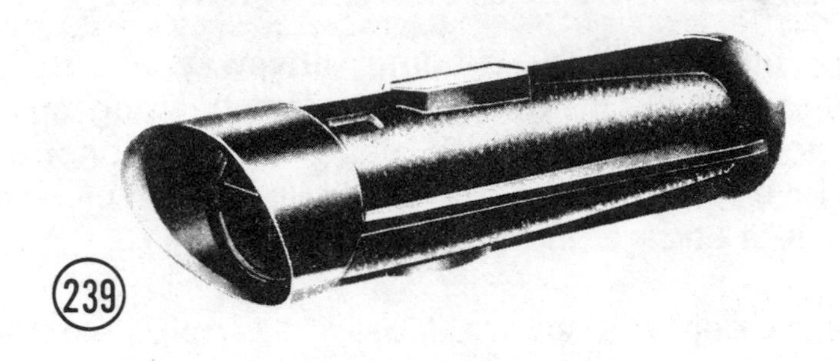

Guide lines that are square, level, or pitch to grade job requires, are easy to establish, and are a must for every construction job. You can set up guide lines in a number of different ways. One accepted way is explained on page 24.

To stake out a site, erect batter boards, check footings, foundation forms, or lay out a terrace with the proper angle of slope to provide drainage, use a level, level-transit or sight level.

ESTABLISHING ACCURATE GUIDE LINES

During the earliest days of civilized man, some smarty figured out that the most accurate distance measurement required a perfectly straight line of sight. Today's modern level and level-transit work on the same basic principle. Since a line of sight is a perfectly straight line, every point along the line is level with all other points on that line. If you are building a wall and want to make certain the guide lines, footing forms, or any course is level, set up a level-transit and check the wall at various points.

When starting any job, measure the exact distance from house or property line to one corner of the project. Drive a stake flush into the ground at point selected. Drive a 4 or 6 penny nail, or tack, into the top of the stake to indicate the exact corner.

To set up a level or level-transit spread the tripod legs at least 36 inches.

Visually set the tripod so the horizontal circle #3, Illus. 237, is level and centered over the nail. Press the legs into the earth to make certain the tripod is steady. Hang a plumb bob from the tripod so point of bob is centered directly over nail, Illus. 240.

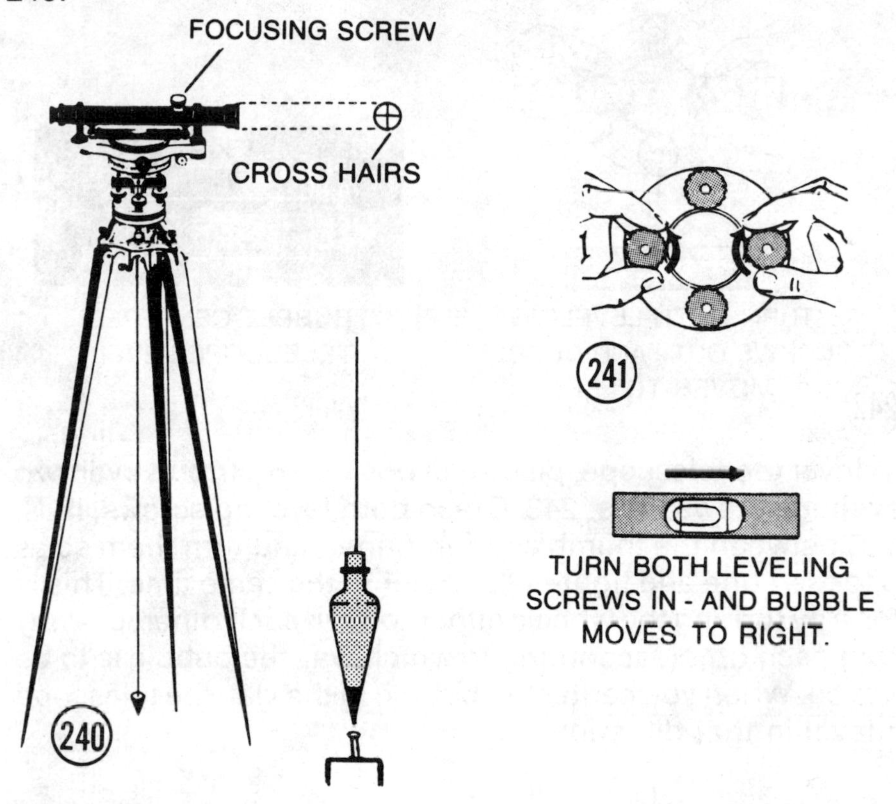

After leveling the instrument as much as the tripod allows, start to "fine tune" by turning the leveling screws #8, Illus. 237, 241.

To fully appreciate how a level, Illus. 237; or a transit-level, Illus. 238, functions, note the various adjustments. The telescope itself is a precision made sighting device with a carefully ground and polished lens that produces a clear, sharp, magnified image. The magnification of a telescope is described as its "power." An 18 power telescope will make a distant object appear 18 times closer than when viewed with the naked eye. Cross hairs on telescope permit the object sighted to be centered exactly.

The leveling vial, Illus. 242, also called a bubble, works like any familiar carpenter's level, but on the transit-level it is more sensitive and more accurately mounted. Four leveling screws, #8, Illus. 237, permit leveling the vial perfectly.

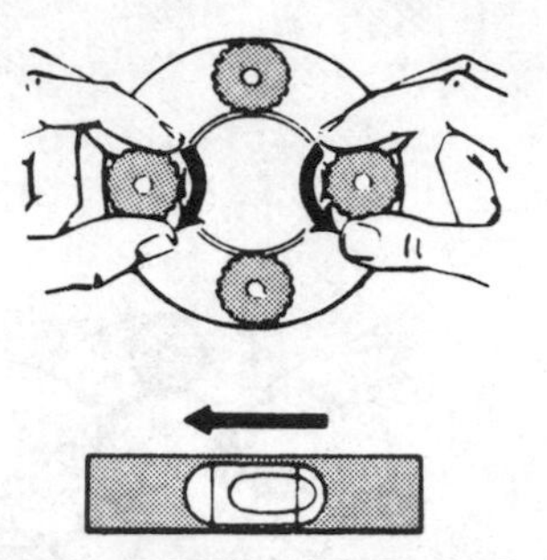

TURN BOTH LEVELING SCREWS OUT - AND BUBBLE MOVES TO LEFT.

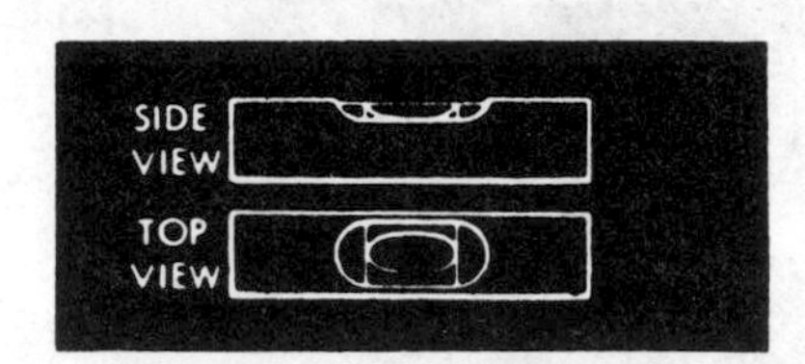

BUBBLE CENTERED TELESCOPE LEVEL

242

To level the telescope, place it in position on tripod over two leveling screws, Illus. 243. Grasp both leveling screws, Illus. 241, between the thumb and forefinger, and turn them so as to loosen one and tighten the other at the same time. This is done by turning the wheels either toward each other, or away from each other, according to which way the bubble is to be moved. When you center the bubble in the vial, the telescope is level in that direction.

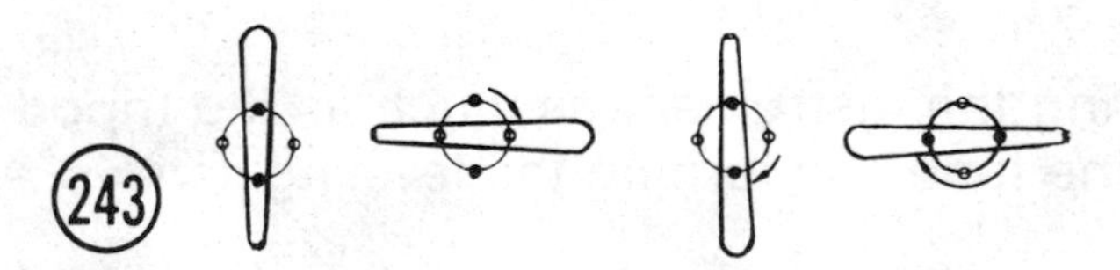

243

Now turn the telescope at right angles(90°), and center it over the other pair of leveling screws. When level in this direction, it is considered level in both directions. Don't tighten the leveling screws too tight, or leave them too loose. If loose, errors will occur; if too tight, the instrument will be strained.

If leveling the instrument has thrown the point of bob off the nail, correct the position of the instrument. Again, check to make certain it's level. To focus, turn knob #2, Illus. 237.

The circle, Illus. 237, is merely a flat plate on which the telescope rotates. It is marked in degrees so the telescope can be rotated in any horizontal direction, and any horizontal angle can be quickly measured on this circle. Most level-transits have a vernier scale #4, Illus. 237, which divides each degree into minutes for accuracy. There are 60 minutes in each degree — 360° in a complete circle.

When the instrument is level, we know the line of sight is perfectly straight. Any point on the line of sight would be exactly level with any other point.

Illus. 244 shows how you can check the difference in height (or elevation) between two points. With an assistant holding a leveling rod, Illus. 245, or a folding rule, Illus. 1, make a test.

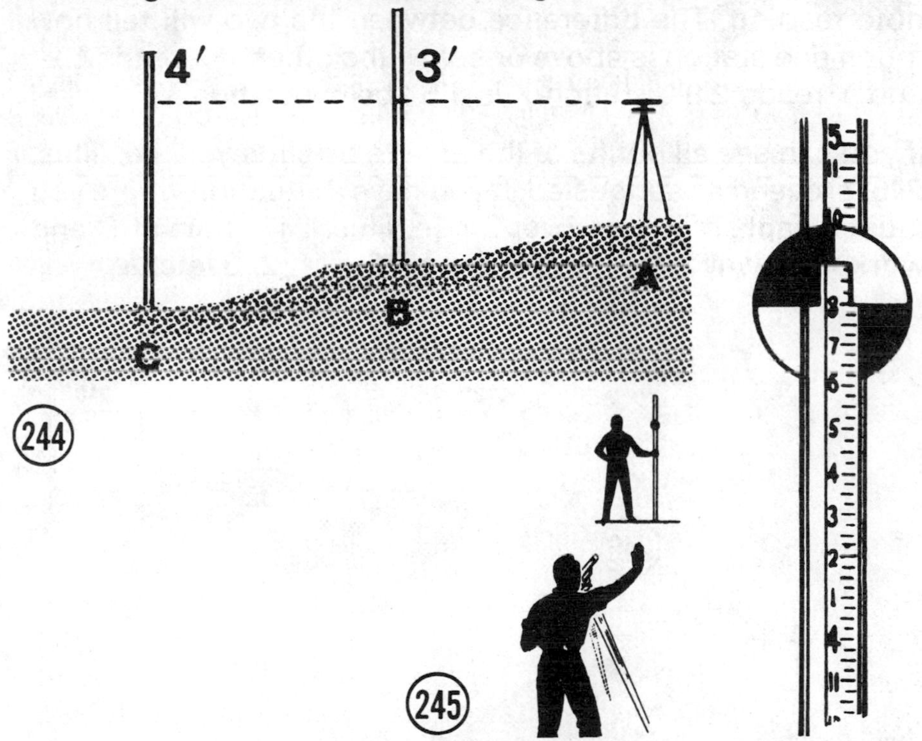

When the rule reads three feet at B, and four feet at C, Illus. 244, you know there's a difference of one foot. Using the same principle, you can easily check to see if a footing form is level; if concrete blocks, a course of brick, windows, or doors are in line. You can also see how much a driveway slopes.

If you decide to purchase a level-transit, be sure to obtain one with the zoom lens and the magnification you need for the type of work you plan on doing.

When sighting through the telescope, keep both eyes open. You will find this eliminates squinting, doesn't tire the eyes, and gives the best view through the telescope.

To establish the difference in grade between several points, do this. Set the instrument about midway between two points. Level the instrument as previously explained. Hold the rod straight up at X. Read the telescope, Illus. 245, 246. When the horizontal cross hair lens cuts the graduations on the rod, note the reading. Place rod at Y. Without disturbing the instrument, swing the telescope clear around, sight rod, note reading. The difference between the two will tell how much one station is above or below the other. If X reads 4'2" and Y reads 2'8", station X is 1'6" below station Y.

If you can see all points of the area to be surveyed, i.e., Illus. 246, proceed as suggested. If you have a situation where you must establish grades over a rise, Illus. 247, start at C and work your way over the hill. Do section 1, 2, 3, etc.

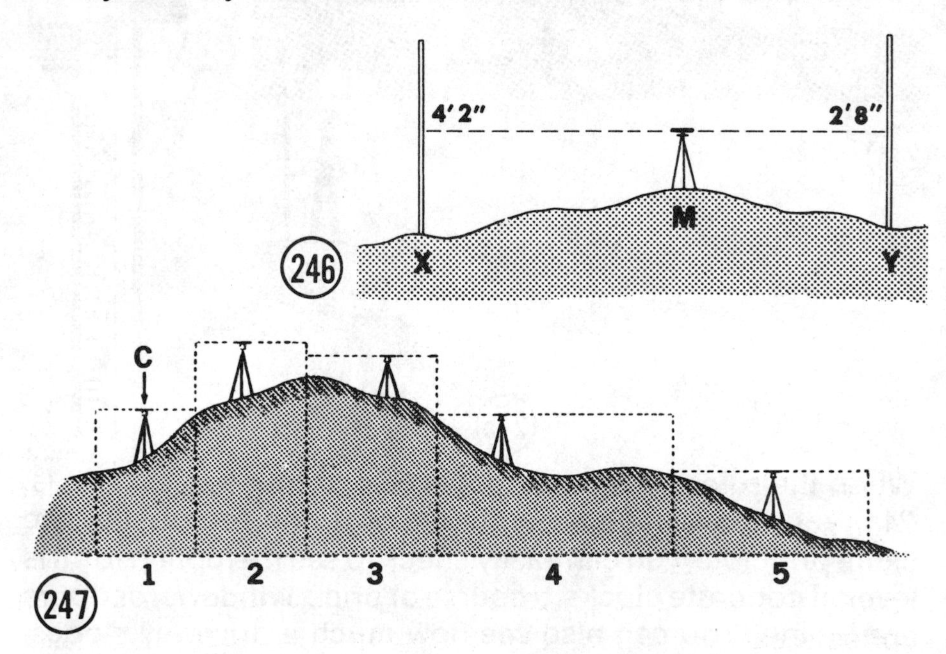

To lay out a site for an addition, garage or other building, measure the distance from the house or property line, and drive a 2 x 2 stake flush into the ground to indicate one corner. Drive a tack or 4 penny nails into the stake. The nail now represents the exact corner. Place tripod in position making certain it's pressed firmly into ground. If you are working on a paved surface, be sure the points on tripod are spread 3½'. Center plumb bob directly over the nail.

Focus along line AB to establish the front or one side of the building, Illus. 248. Using a steel tape, measure exact distance required, drive a stake, and nail into top of stake at exact corner. To double check position of nail, hold a plumb bob over the nail. Sight line of plumb bob. To establish line AD, bring the telescope 90° on the circle scale #3, Illus. 237. Measure distance of AD, drive a stake and a nail into top of stake.

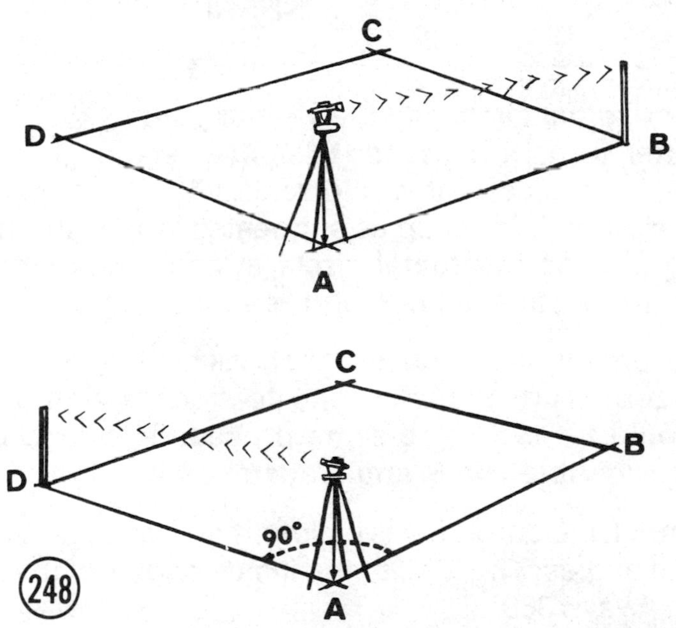

To establish CD and CB, set the tripod up over D. Sight to A and set circle at zero. Turn telescope 90° to establish line DC. Measure distance required and drive a stake at C. Follow same procedure to establish CB. Any number of additional offsets can be laid following the same procedure.

The plumb bob over nail not only indicates a corner, but also the zero point of each angle. To establish corners on an odd shaped site, Illus. 249, set the instrument up at station 1, level it as explained previously, swing the telescope to establish station 2.

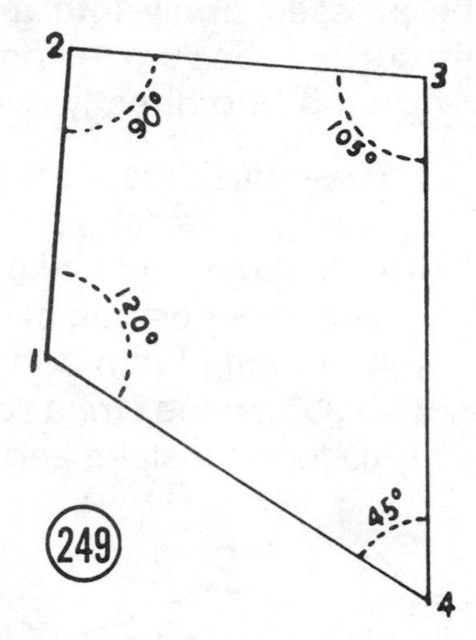

Set the horizontal circle and the vernier guide #4, Illus. 237, at zero. Then turn the telescope to sight on station 4 and read the angle. In this case, it would read 120°. Next move the tripod to station 2. Level up as suggested, and sight to read station 1. Set the horizontal circle at zero, then sight the telescope to locate station 3, and read the angle 90°.

Place instrument on station 3, sight back to station 2. Set the circle at zero. Turn the telescope to sight station 4. Your angle should be 105°. The same procedure is followed to measure or double check angle at station 4.

Double check the correctness of your readings by adding the four inside angles. The total of the inside angles of the quadrangle is always 360°.

To lay out any angle, proceed in the same way. Set the instrument at station 1, and set the circle at zero. Swing the telescope to the desired angle. Place the rod in position so that it intersects the vertical cross hairs. Drive a stake. One leg of an angle is thus established.

RUNNING STRAIGHT LINES WITH A LEVEL

Level the instrument over the first point, Illus. 250. Hold the rod at the first stake point and adjust the position of the target so it falls on the cross hairs of the instrument. Make a note of the height.

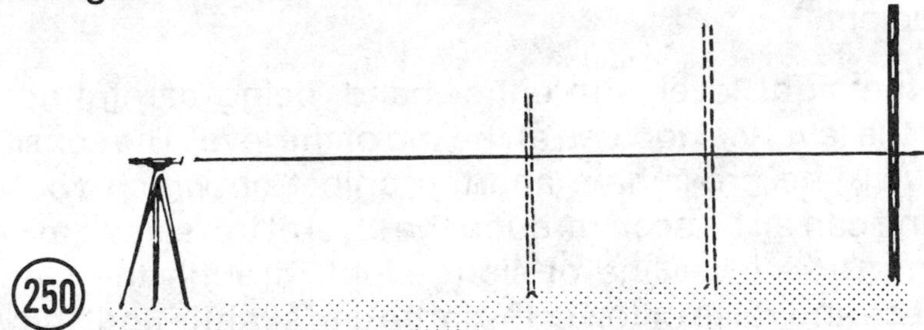

Move the rod to the next point and proceed as before. Repeat as many times as necessary. When the rod moves beyond the range of the instrument, set it up over the next to the last mark point. Focus the instrument until the target falls in the cross hairs. Proceed as before. This is an invaluable aid in setting up fencing so the top rail is level all the way through.

ESTABLISH VERTICAL LINES AND PLANES

To establish vertical lines and planes, use a level-transit. Level the instrument as previously described, then release the locking levers which hold the telescope in level position. Swing the telescope vertically and horizontally until the line to be established is directly on the vertical cross hair. If the telescope is now rotated up or down, Illus. 251, each point cut by the vertical cross hair is in a vertical plane with the starting point.

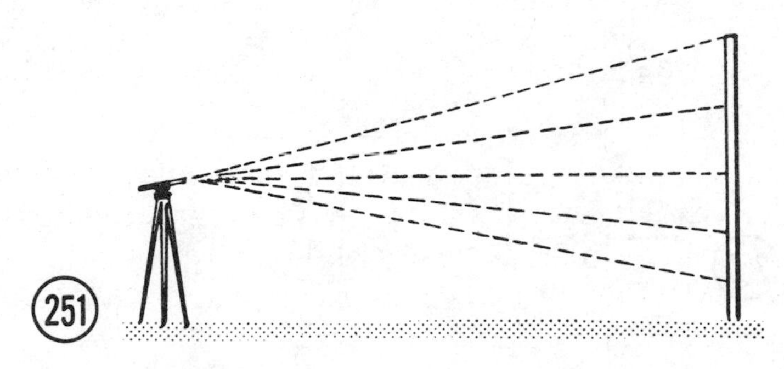

SIGHT LEVEL

A sight level, Illus. 239, is a carefully made precision instrument designed to simplify many jobs. It has a spirit level and cross hair that indicate true level of sight. It was developed using the same principle of operation as a surveying instrument.

Hold the sight level with either hand, being careful not to cover the level vial located at the top of the level. The sensitive level vial, the cross hair, and the object on which you are sighting can all be seen through the level at the same time. To obtain a true level line of sight, sight through the slotted eyepiece and raise or lower the large end slightly until the vial is level and centered with the cross hair. A level line of sight is a continuous, perfectly straight line for as far as you can see. All points and objects along this line are exactly level with your eye.

The sight level permits accurately checking the level of retaining walls, fences, masonry, batter boards, foundations, etc., by simply sighting on the object and noting its position in relation to the level line of sight through the level. To establish or determine differences in level, an assistant holds a measuring rod or folding rule and notes measurements.

Sight through the level with vial bubble centered. Have your assistant note or mark off measurements on the rod. For example, the first reading at A, Illus. 252, is 4'. Have rod moved to location B. Without moving your position, take another reading. If the reading on the rod at B is 5', the difference in elevation between the two points is one foot.

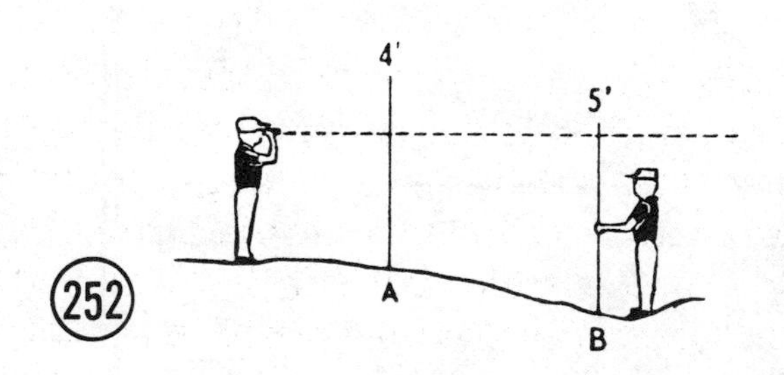

HOW TO ESTABLISH SLOPE OR PITCH

If you want to lay a drain or grade a surface and want, for example, a pitch of ¼" per foot, and the end point is 16', place the measuring rod on selected point and take a reading. If point A, Illus. 253, reads 5'0", and you want a reading of 5'4" at B, excavate to depth required.

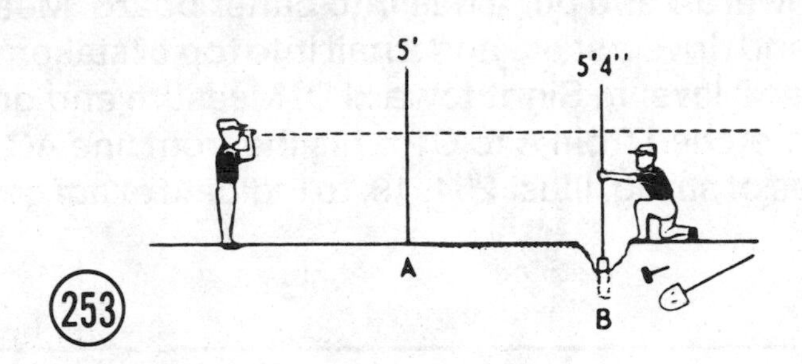

This establishes a 4" pitch in 16'0" and tells you where and how deep to dig. For more accurate readings over long distance, steady the sight level against a board rested on any solid object.

TO LAY OUT A BUILDING AND SET UP BATTER BOARDS

To set up batter boards, Illus. 12, so the top edge is level with all other batter boards, follow this procedure.

Set the transit in the middle of the building site, and level as previously outlined. Decide what height masonry is to be laid, and drive a stake. Use length needed. Drive a nail to establish a grade mark at point above grade chosen for the completed foundation.

Position bottom edge of target rod so it touches grade mark. Move the rod to next stake but don't move the target. Hold rod against next stake. When line in target matches cross hair on lens, mark stake. Drive two more stakes to indicate one corner. Check each stake with the rod. Nail batter board to the stakes so top edge of batter board is at height of the original guide mark.

Drive stakes and nail batter boards at the same level at all corners. When all batter boards check level, drive a nail into top edge of batter board to establish location for building lines. Set the transit over stake A and level it. Sight toward B. Using a plumb bob to guide you, drive a nail into batter board. Measure from A to B and drive a stake to indicate exact corner of B. Set a nail in stake to indicate the exact corner. Now sight toward J and put a nail into batter board. Measure from A to C and drive a stake and a nail into top of stake. Move transit to B and level it. Sight toward D. Measure and drive a nail. A line stretched from A to C marks the front line AC. You can tie pieces of string, Illus. 254, 19, to indicate exact corner.

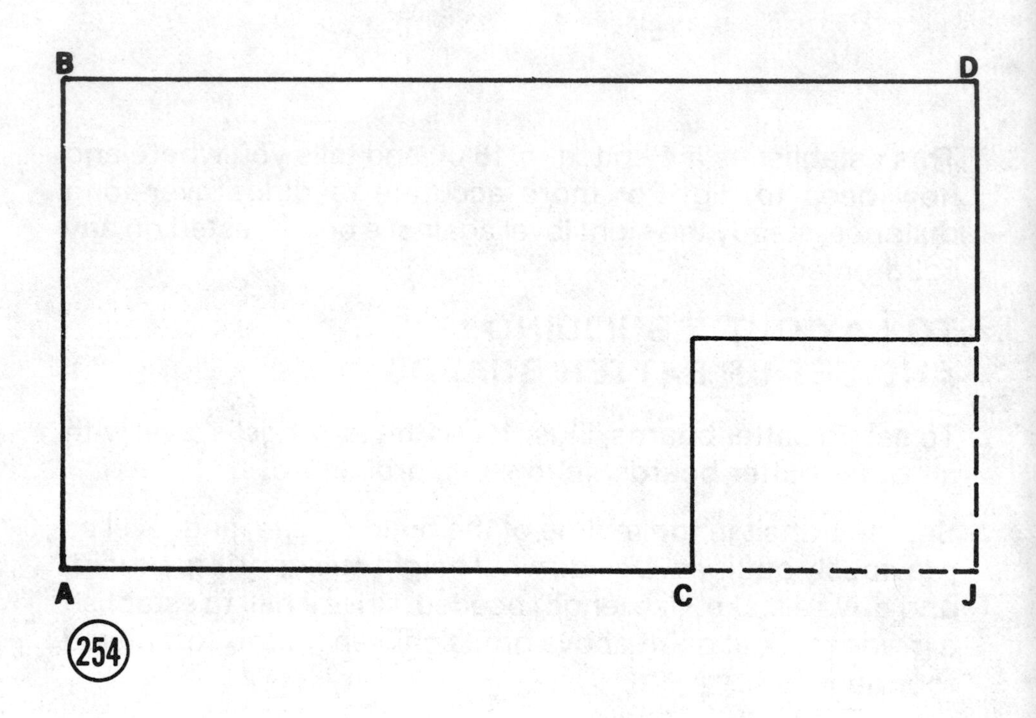

STAIR BUILDING — INTERIOR

Considerable money can be saved building stairs on the job. Those who need to save time should price precut components. If you decide to buy rather than build, ask the retailer to check location and take needed dimensions. Whether you purchase precut components or assembled stairs, frame the opening prior to delivery. Retailers frequently send extra help when delivering assembled stairs. This permits nailing in place.

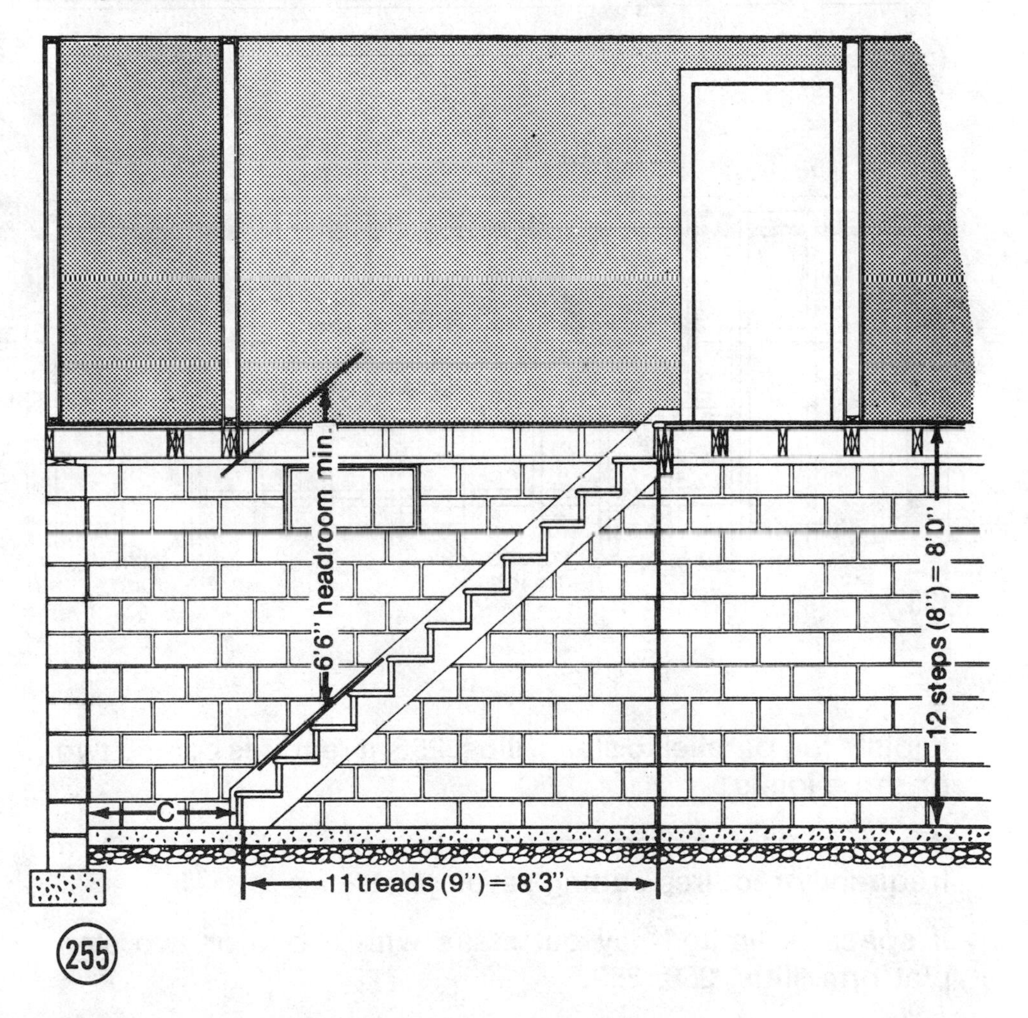

255

Select space for stairs where distance C, Illus. 255, is 36" or more from wall.

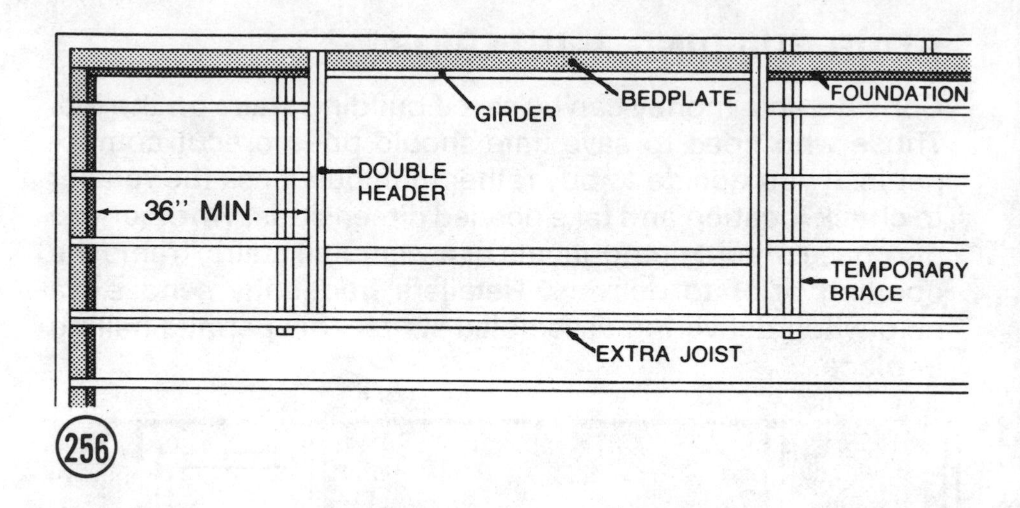

256

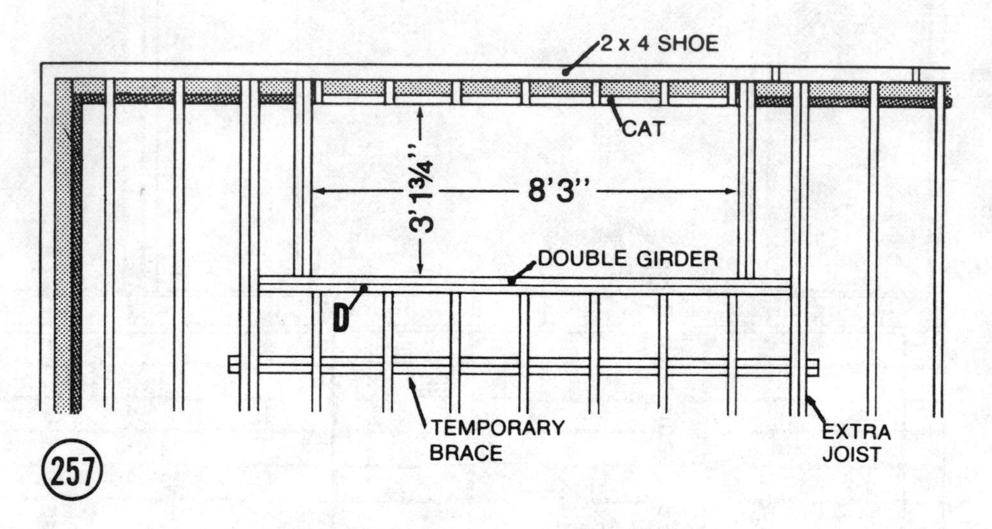

257

If joists run parallel to stairs, Illus. 256, it requires cutting two or three joists.

When joists run perpendicular to opening, Illus. 257, it frequently requires cutting seven joists.

If space is limited, lay out stairs with a one or two step platform, Illus. 258, 259.

Winder stairs, Illus. 260, and prefabricated spiral stairs, Illus. 261, provide two other solutions.

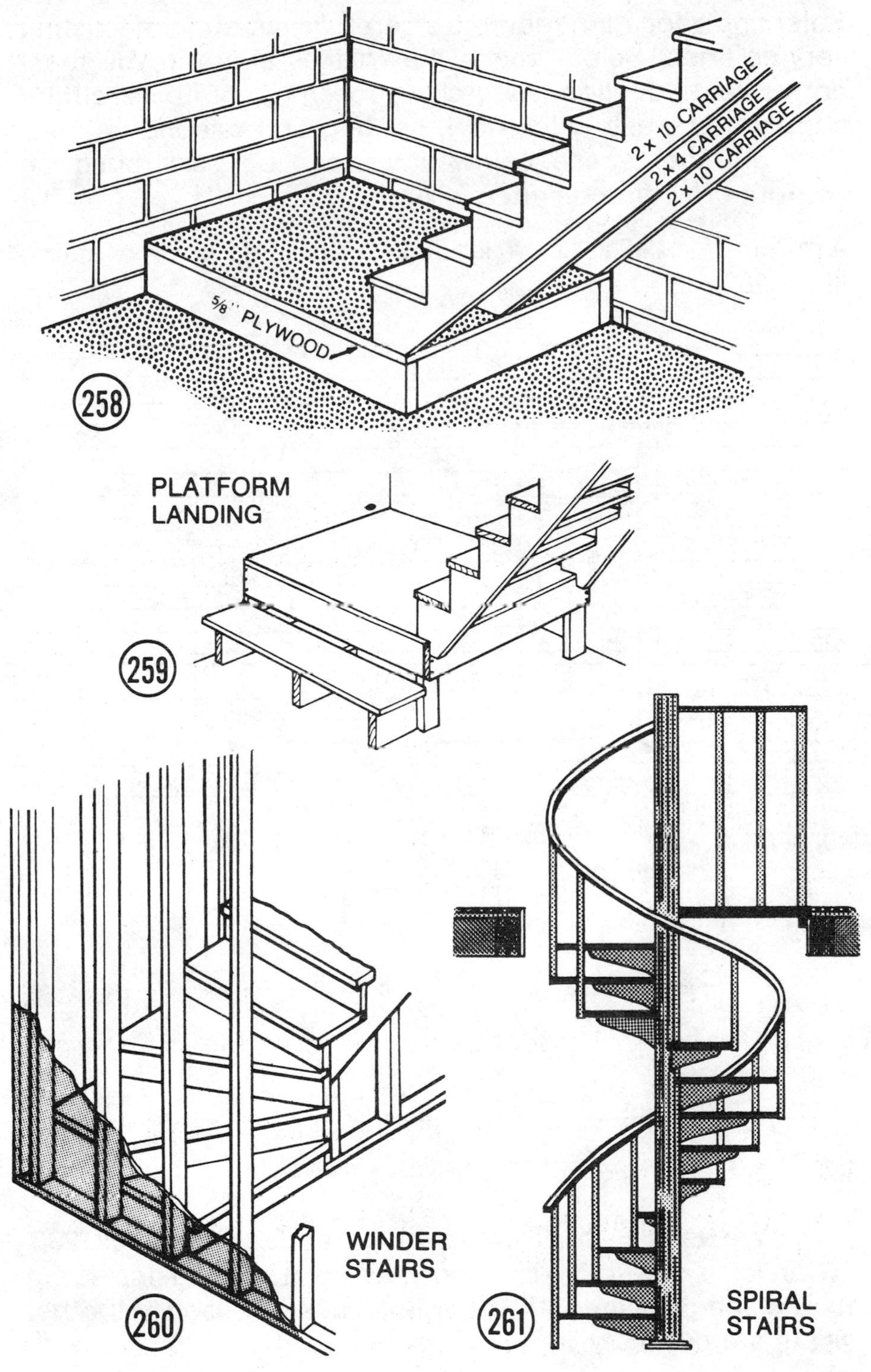
2 x 10 CARRIAGE
2 x 4 CARRIAGE
2 x 10 CARRIAGE
5/8'' PLYWOOD
258
PLATFORM
LANDING
259
WINDER
STAIRS
260
261
SPIRAL
STAIRS

Selecting a location requires a careful evaluation of existing service lines. Select space that places stairs in the most convenient location. One that doesn't require rerouting supply and waste, electrical, heating and telephone lines. Note whether space below stairs can be partitioned off without crowding balance of area.

A minimum of 42 x 42" is needed for a landing at top of stairs, Illus. 262.

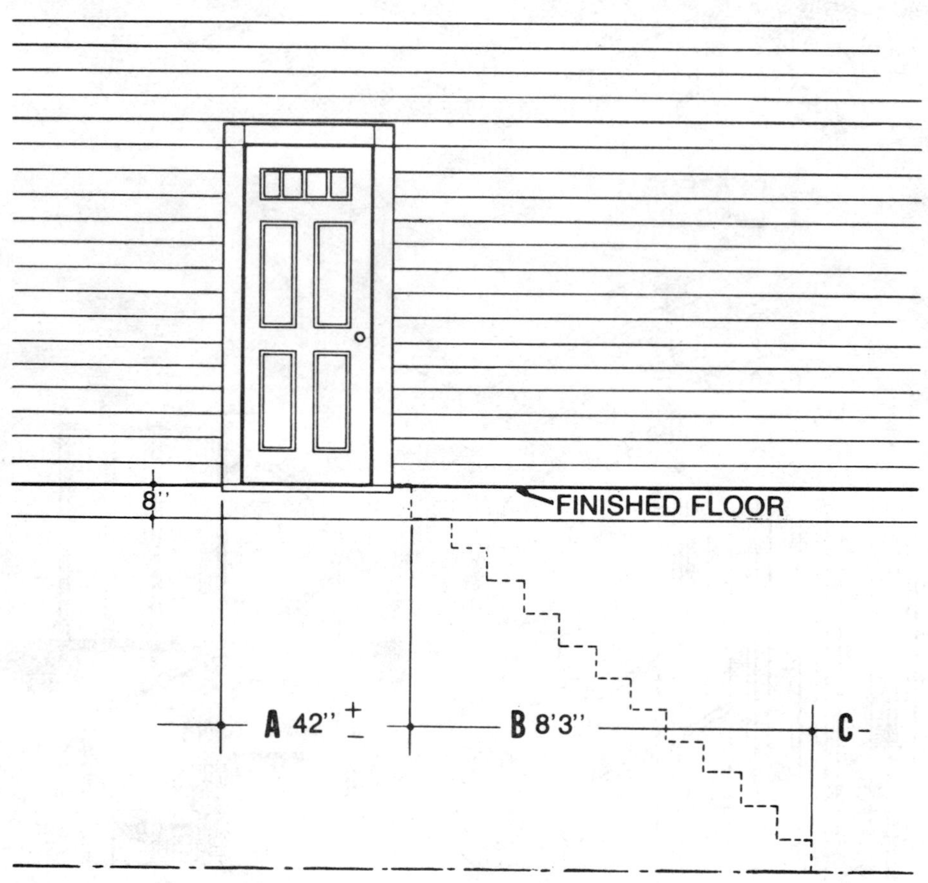

262

Stairs are usually constructed with three carriages, Illus. 263. When laid out with 8" riser and 9" tread, a starting riser is cut to 6⅞" height when 1⅛" inside treads are used. All other risers are cut 8".

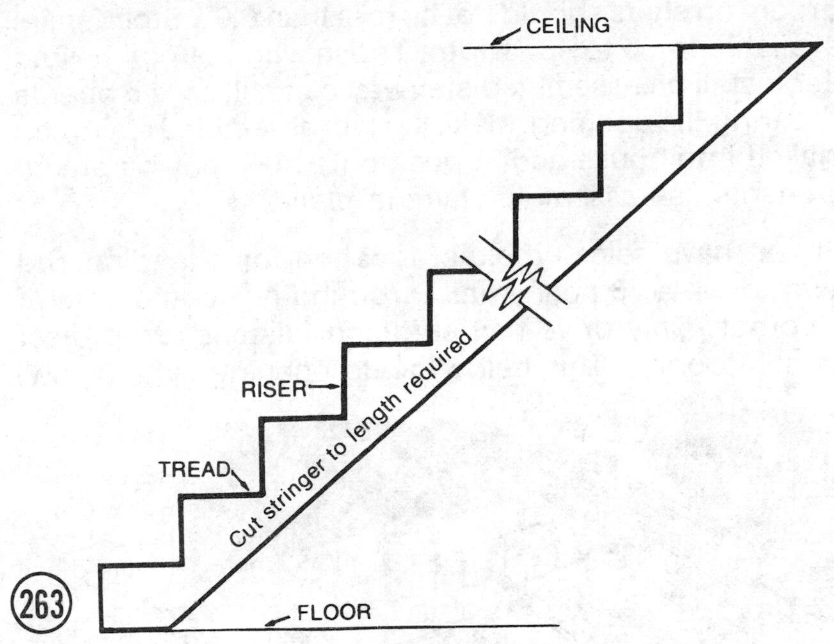

If overall height can't be divided by 8, try 7¾, 7⅝ or 7½", or absorb any measurement less than 8" with a starting platform.

To ascertain number of risers and width of tread, the unskilled craftsman uses an equation that specifies 2R + 1T =25. (R = riser, T = tread). To appreciate how this works, drop a plumb bob down from floor to floor, Illus. 255. Measure exact distance and divide by 8. Use 8" because this is the maximum height specified for risers. If overall height measures 96", divide by 8 = 12. This means you will need 12 equaly spaced steps.

Codes frequently specify Riser x Tread = 72 to 75. An 8" riser x 9" tread = 72. If 7¾" (riser) divides equally into overall height, use a 9½" tread: 7¾ x 9½ = 73+ which is OK. Where codes specify 2R + 1T = 25, two 8" risers, plus one 9" tread equals 25. If you need to change risers to 7¾"; 2R 15½ + 1T 9½ = 25.

To determine width of tread, add two R, subtract from 25 and you have Y—width of tread. If R = 8, two R = 16. This indicates a 9" wide tread.

Headroom on stairs should not be less than 6'6''. Stock stairs are available in 30 to 48'' width. Those who plan on having retailer install an assembled stairway can still save a bundle by preparing the opening. Make certain assembled stairs can be snaked into house and in position. Those buying precut components can assemble stairs in place.

When you have selected exact location for a landing and stairway, drive a 16 penny nail through finished flooring at each corner. Only drive nail sufficient distance to project below subflooring. This helps locate opening from below.

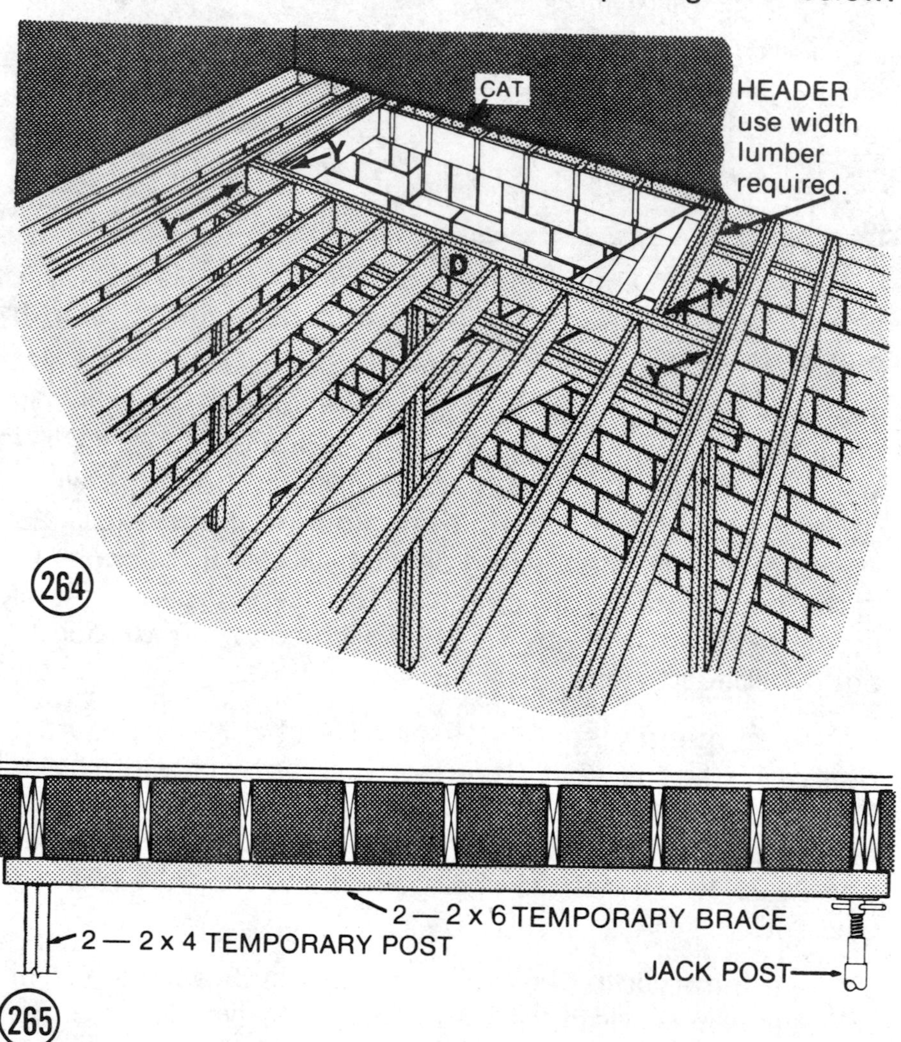

Prior to cutting an opening, support joists with temporary bracing, Illus. 264, 265. Nail two 2 x 6's. Toenail to joists. Support with adjustable screw jacks or double 2 x 4 posts every six feet. Keep bracing as close to opening as possible without interfering with needed working space.

Drill a hole at each corner of proposed opening, Illus. 266.

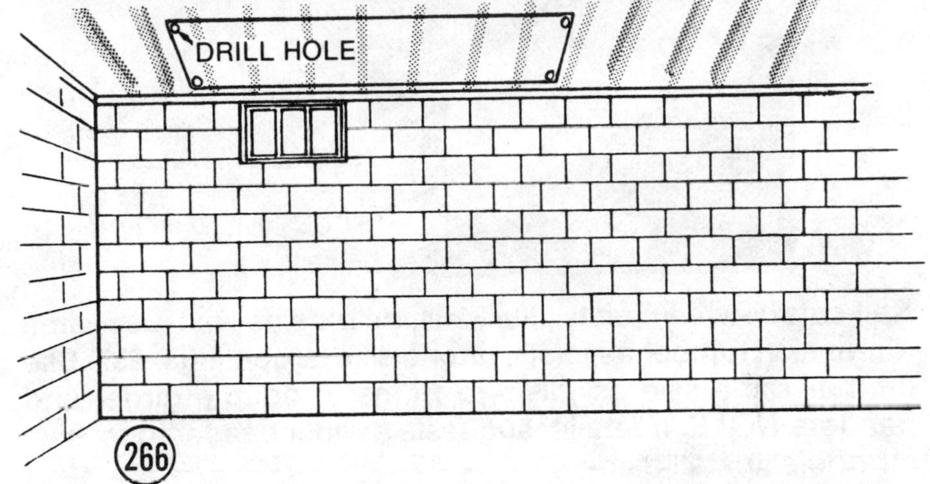

Illus. 267 shows framing needed around stair opening. Note ¾" spacing allowed for 1 x 2 or 1 x 3 furring on left side. This provides nailors for prefinished paneling or plasterboard. Cut opening so sub-flooring covers edge of furring on out side wall; DD and D on inside wall, note Illus. 284.

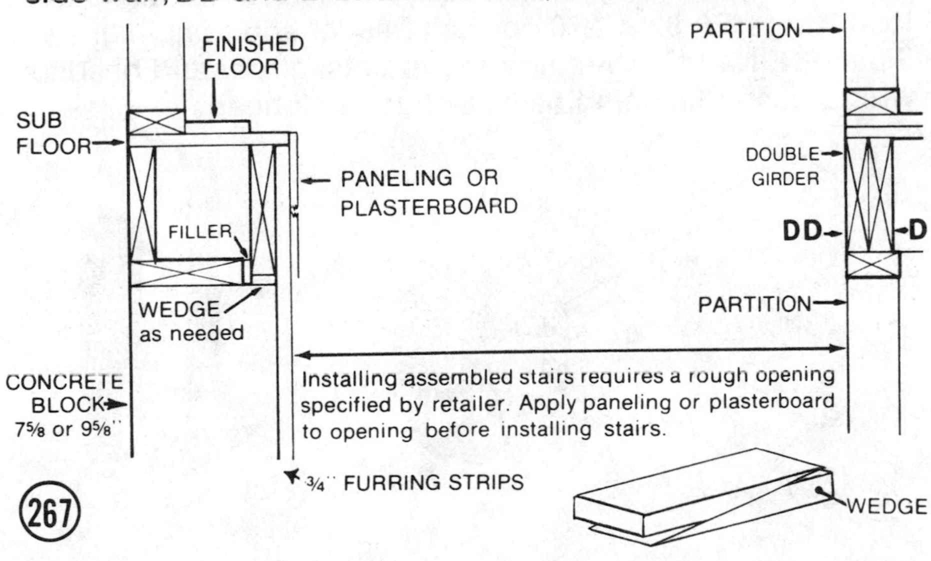

One way to saw joists to length that allows flooring to project amount required is to drill a hole in joist, Illus. 268. Using a level, draw a plumb line. Insert saber or reciprocating saw blade and saw joist up to flooring.

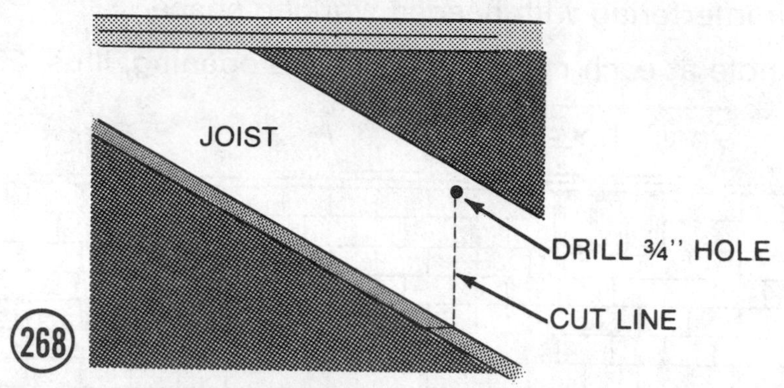

Nail cats between joists, Illus. 264, on outside wall. Use same dimension lumber as joists, or one size larger, Illus. 257. Use lumber same size as joists to frame in double girder and headers. NOTE: If retailer suggests a wider header to receive stringers, install same.

Toenail cats to joists. Nail through D, Illus. 257, into ends of cut joists; through joist into ends of D. Spike DD in place. Toenail headers to girder. Insert extra joists where shown.

Nail single joist hanger X to each joist, Illus. 269; nail double joist hanger Y, Illus. 270, to each header and double girder, Illus. 264. Nail finished flooring to cats, girder and headers with 8 penny finishing nails. Countersink heads.

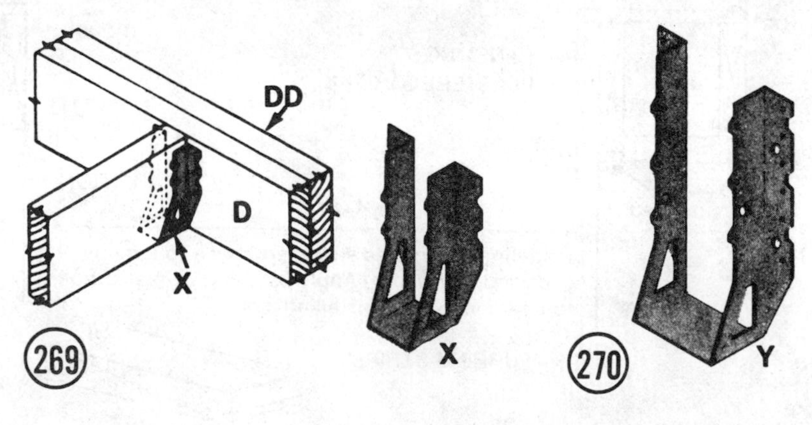

Apply 1 x 2 furring to outside wall, Illus. 267. Space furring 16" on centers. This permits applying paneling to stairwell with adhesive or nails prior to installing stairs.

If stairs require a one or two step platform, Illus. 271, 258, 259, you can use 2 x 4, 2 x 6, 2 x 8 for framing. 2 x 4 on edge covered with ⅝" plywood provides a 4⅛" platform; 2 x 6 measures 6⅛"; 2 x 8 - 7⅞". When a two step platform is required, cut 2 x 4 legs to height required, Illus. 272, 259. Cut 2 x 6 or 2 x 8 frame to size space permits. Reinforce legs with scrap pieces of 2 x 4 nailed to 2 x 8 and to leg. The first tread is nailed to outriggers. The platform is covered with ⅝" plywood.

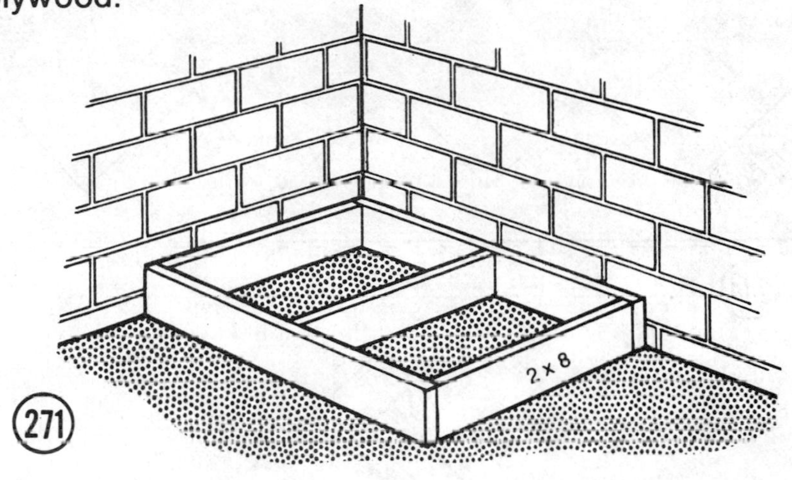

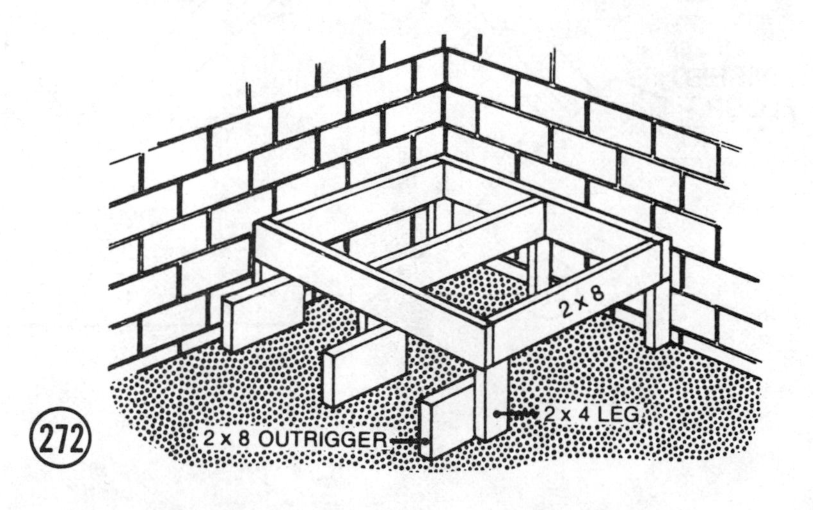

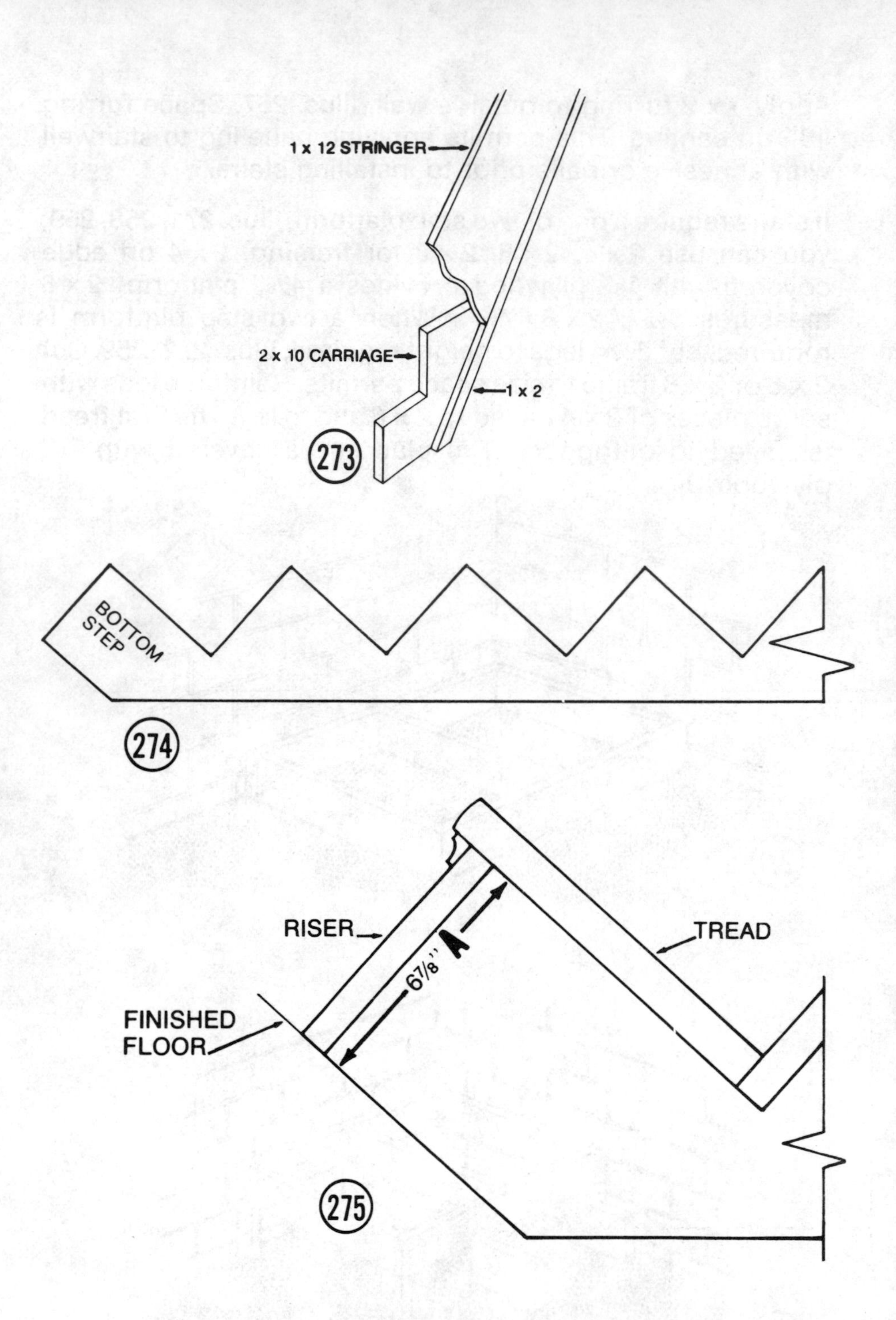
1 x 12 STRINGER
2 x 10 CARRIAGE
1 x 2
273
BOTTOM STEP
274
RISER
TREAD
6 7/8"
FINISHED FLOOR
275

Use a 1 x 12 or 5/4 x 12'' for stringers; 2 x 10 carriages, Illus. 273, 274. If you need a third carriage, use a 2 x 4 and nail cutouts from 2 x 10 carriage in position needed.

To double check overall length needed for stringer and carriage, Illus. 274, cut bottom angle on a 2 x 4 x 6', top angle on a second 2 x 4. Stretch these out so they fit opening. Mark where they overlap. Measure overall length needed.

Stock stair treads and risers, Illus. 275, are routed as shown. Apply glue before assembly.

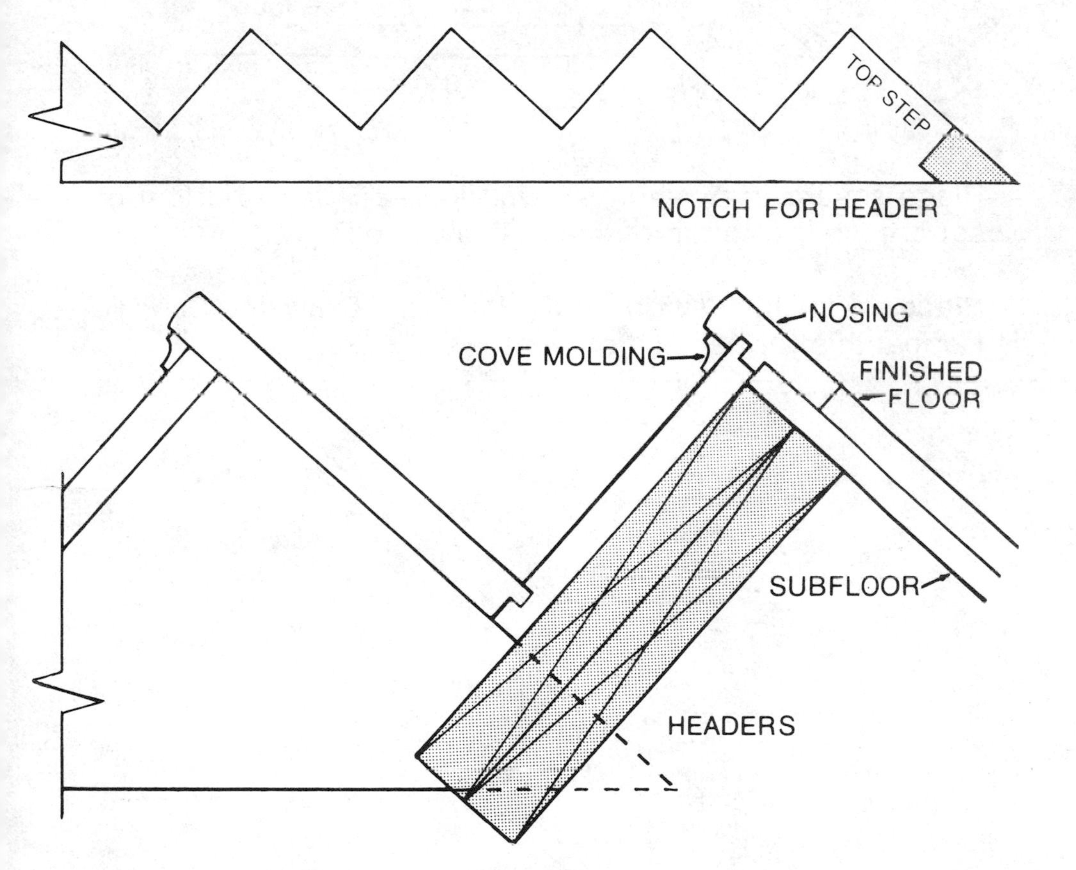

To lay out carriages having an 8'' riser, 9'' tread, place square in position shown, Illus. 276. Draw line A.

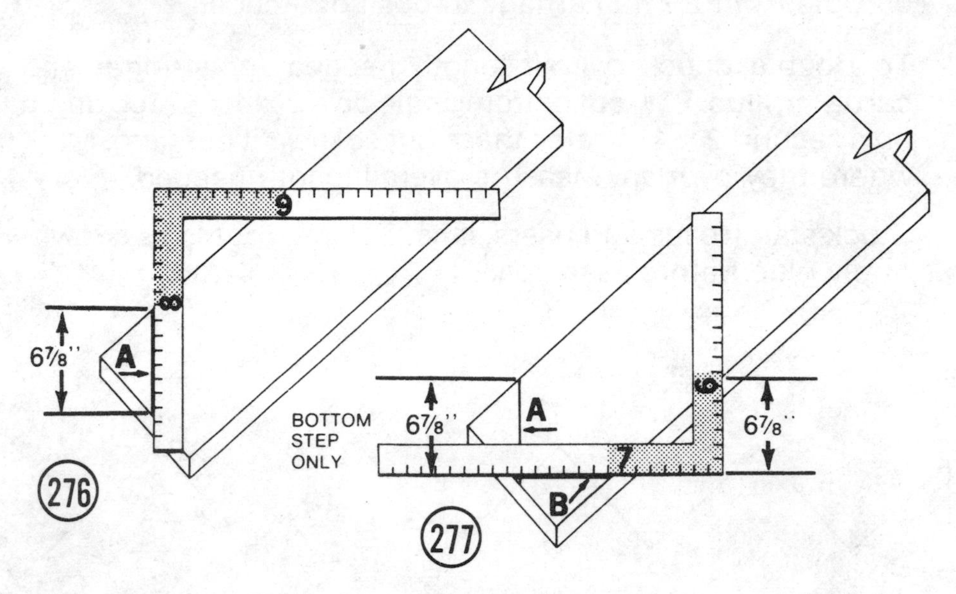

Place square in position shown, Illus. 277, with bottom of square 6⅞'' from top of line A. Draw line B.

Place square in position shown, Illus. 278. Draw line T and R. Repeat this for balance of steps.

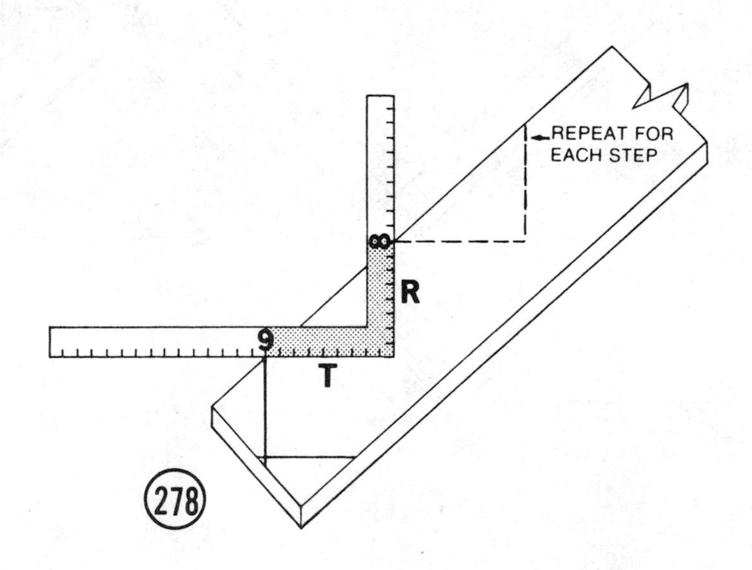

Illus. 262 shows 12 risers, 11 treads. Most millwork houses sell oak treads up to 10½", and will cut these to width required. Buy precut treads and risers to size required.

Nail stringer to carriage, then to header, Illus. 264. Nail cutouts from 2 x 10 carriage to 2 x 4, Illus. 258. Use this for a middle carriage. This provides more support for stairs. It also provides a nailor for gypsum board on ceiling under stairs.

Build partition to length desired, Illus. 279. Since you want to move furniture down stairs, consider how many steps you will want exposed. Spike stringer to studs.

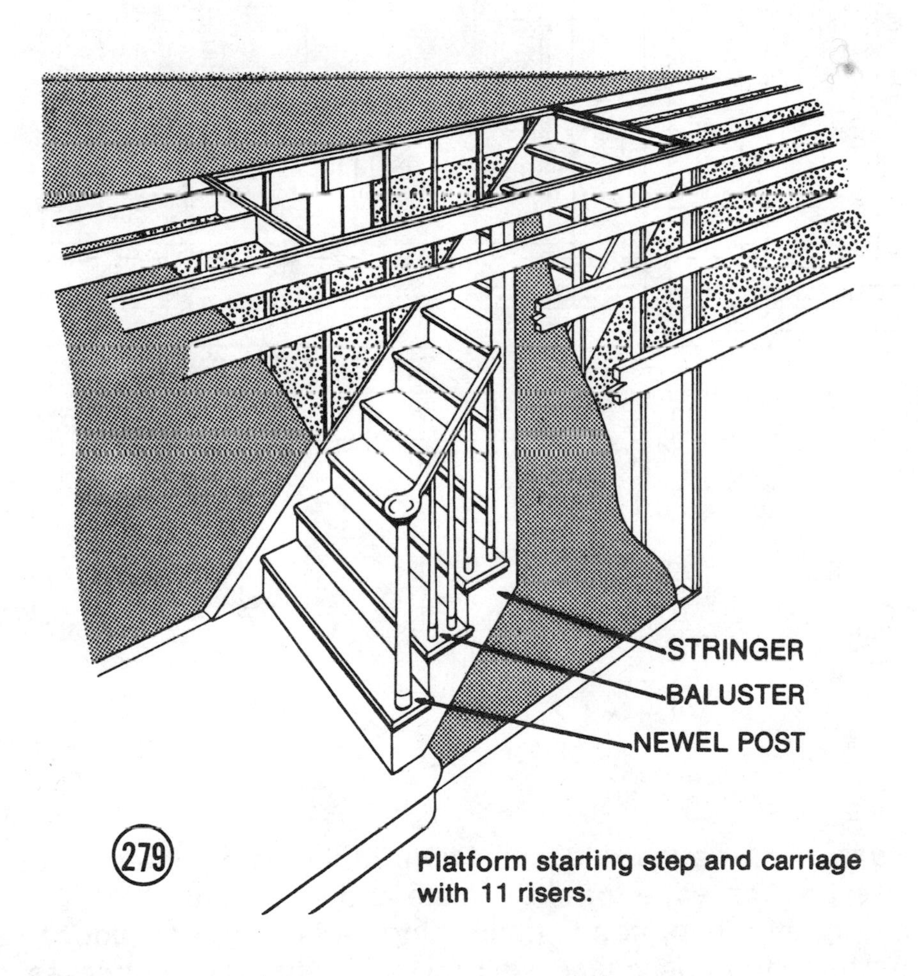

Platform starting step and carriage with 11 risers.

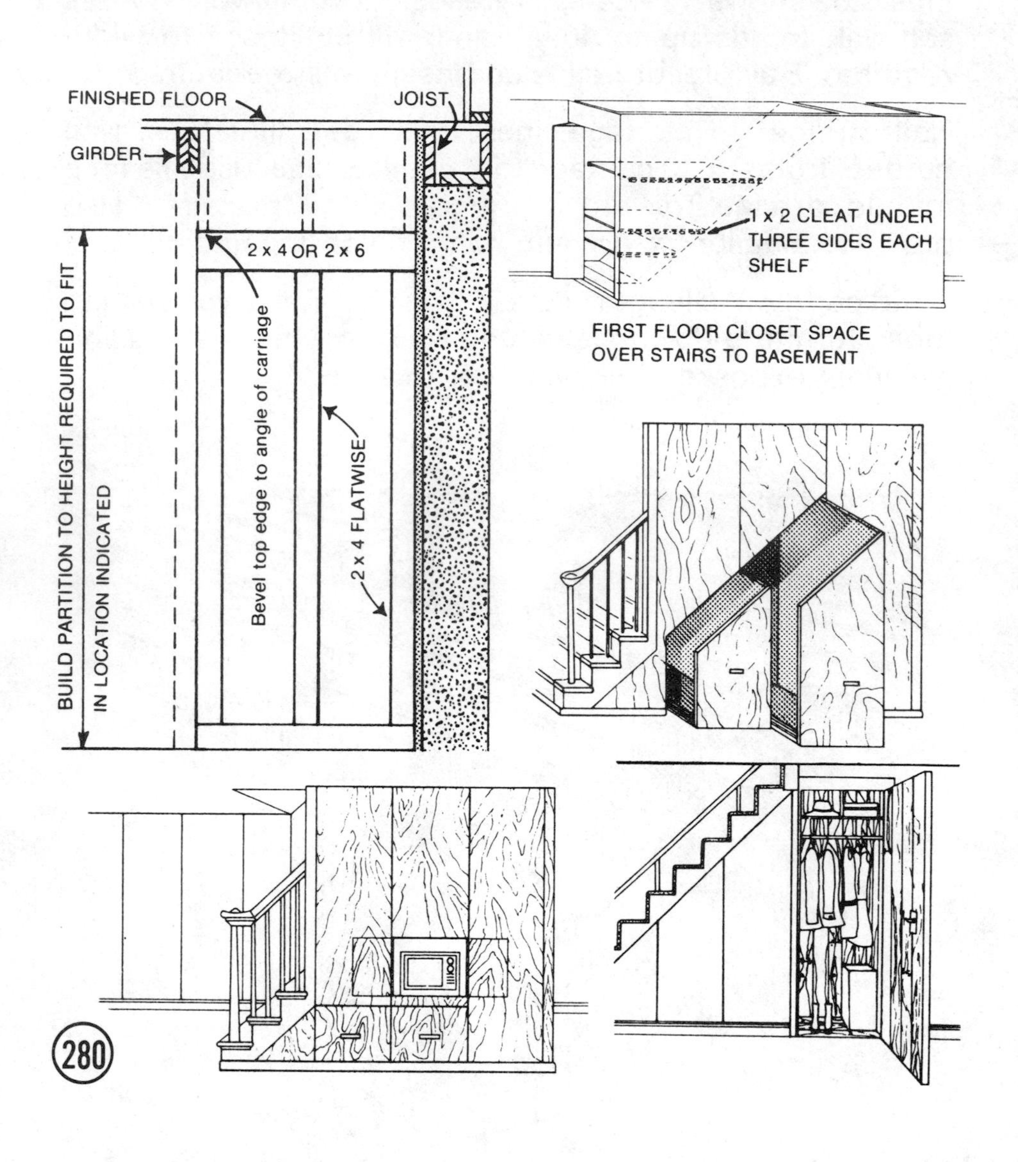

Support carriages with a partition. Build to height required. Use 2 x 4 on edge for shoe, plate and studs, Illus. 280. The center stud is optional. If you hinge a ¾'' plywood door to partition framing, space can provide a catchall for suitcases, skis, or be used for storage.

Apply glue to riser where it butts carriage, Illus. 275. Nail risers to carriage with 8 penny finishing nails. Countersink heads.

Draw lines on tread to indicate position of carriage. You will have to drill holes through oak treads.* Apply glue where tread butts against carriage and riser and nail in place with 10 penny finishing nails. Nail through riser into edge of tread with 6 penny box nails.

Brad cove molding below tread, Illus. 281.

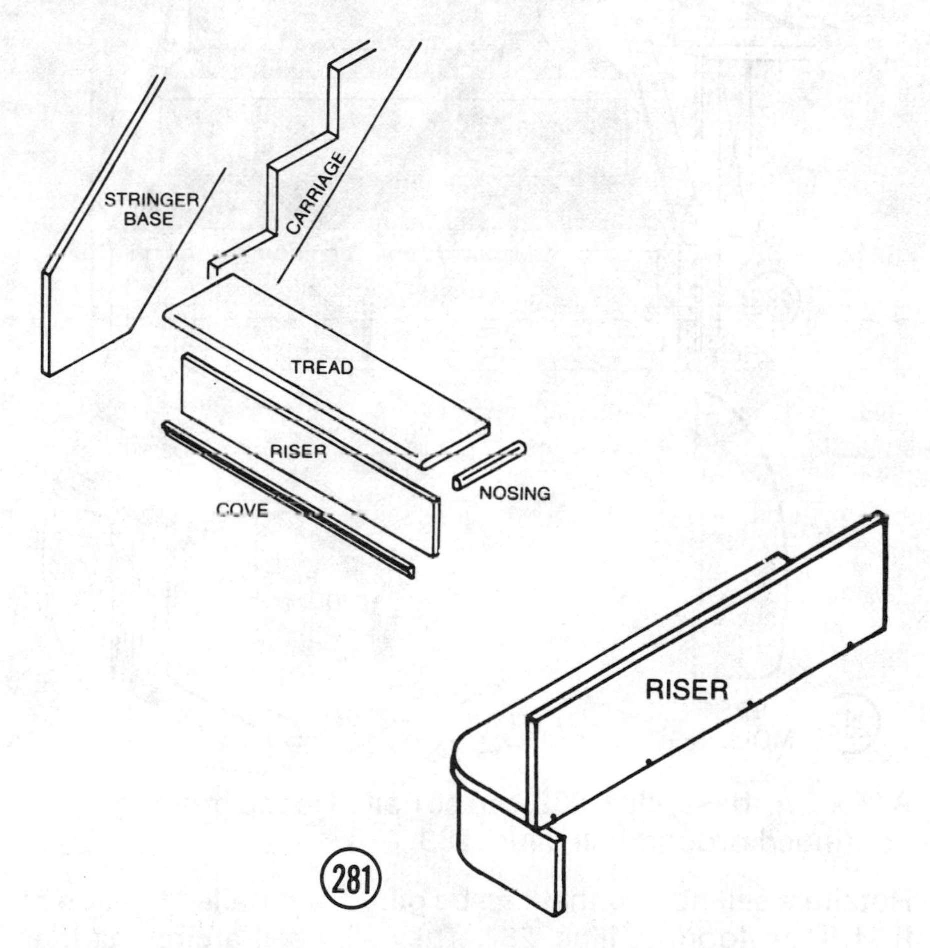

*Some treads can be nailed without drilling.

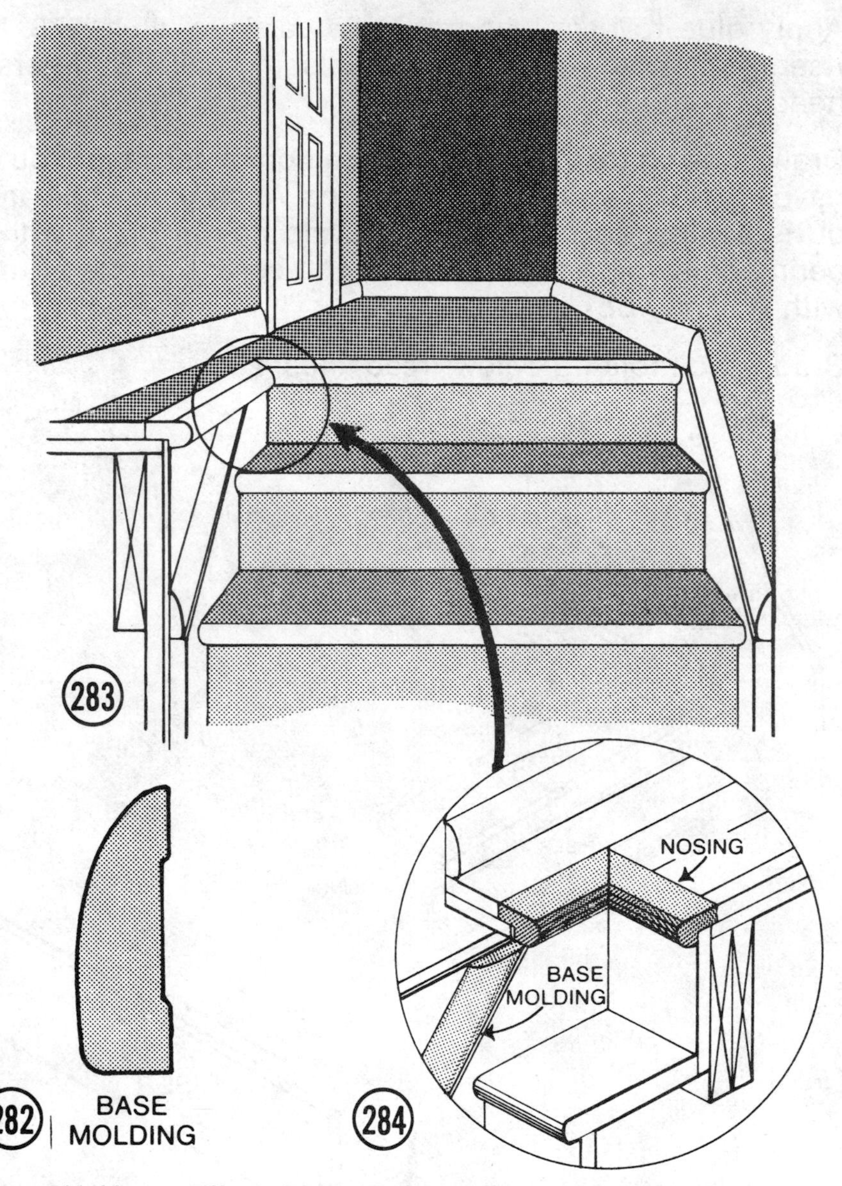

A ½ x 2¼" base, Illus. 282, can be nailed to top of stringer and continued around hall, Illus. 283.

Retailers sell nosing that can be glued and nailed to edge of first floor flooring, Illus. 284. They also sell a circle corner starting step, Illus. 281, 285. Those who buy stair components should find out what's available and space required prior to purchase.

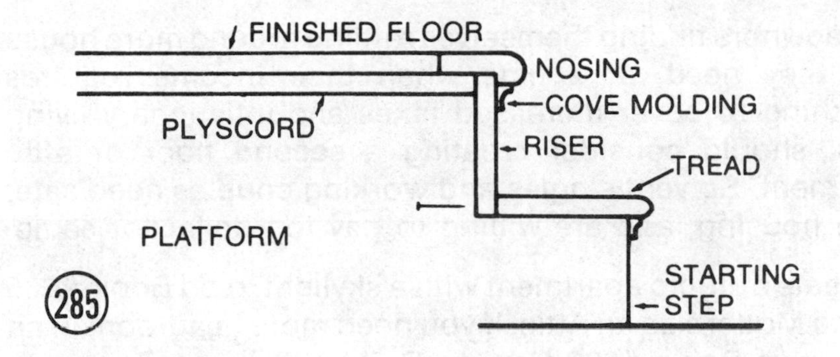

In new construction, nail ¾'' plywood to carriage for temporary steps. If finished treads and risers are installed, cover with building paper until construction is completed.

Build first floor partition to length space requires. Use 2 x 4 for shoe and plate. Space studs 16'' on centers, Illus. 286. Apply prefinished plywood.

Install newel post, balusters and handrail, Illus. 279, at height retailer suggests. These average 2'6'' to 2'8'' above tread. Fasten brackets for handrail into studs.

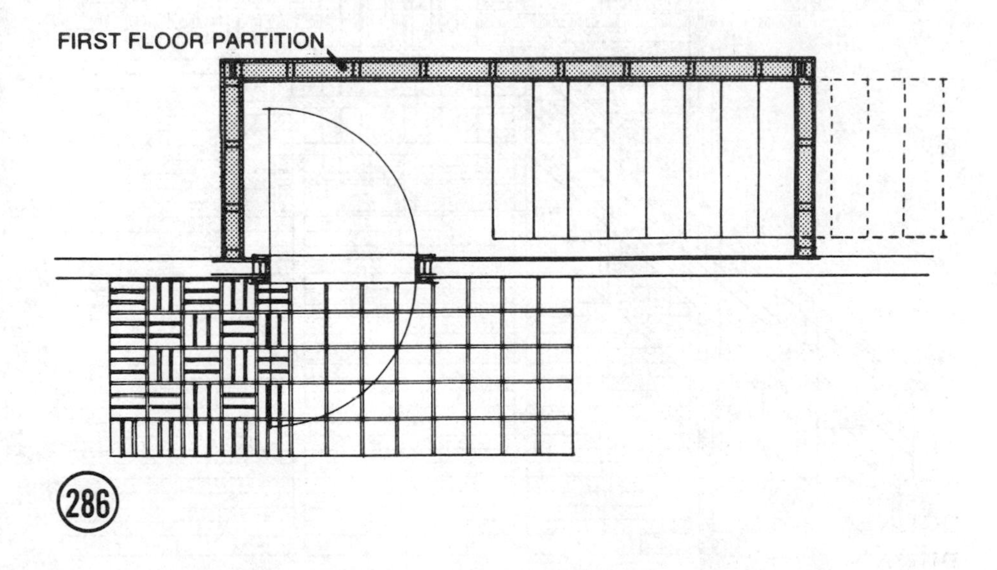

STAIR BUILDING — EXTERIOR

Homeowners finding themselves rattling around more house than they need, at a time when their income requires stretching to cover increased taxes and inflationary living costs, should consider creating a second floor or attic apartment. Solvent singles and working couples need safe, clean housing, and are willing to pay top dollar for same.

To create a studio apartment with a skylight, read Book #665 How to Modernize an Attic. If you need more headroom in an attic, read Book #603 How to Build a Dormer. To create privacy both families need, build outside stairs to a second story.

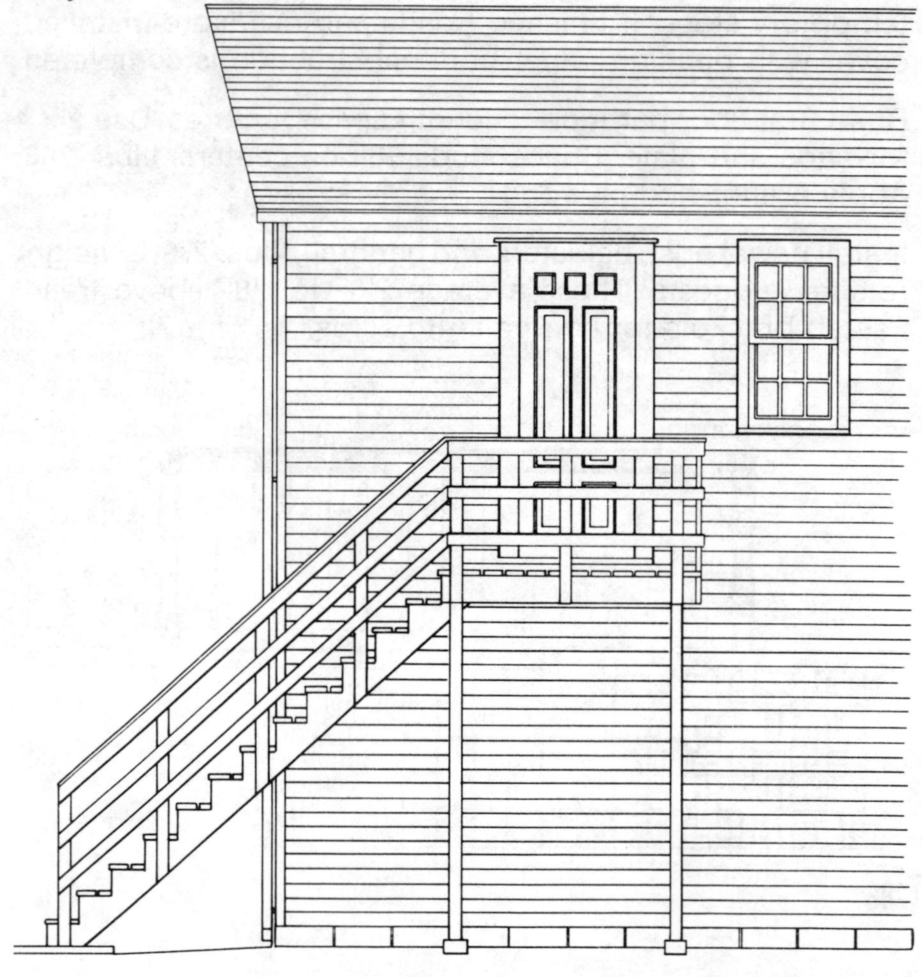

To install outside stairs, measure 4⅞'' down from door opening, draw a level line, Illus. 287. Cut furring blocks A, A1, A2, to length specified. Paint with wood preservative before nailing in place.

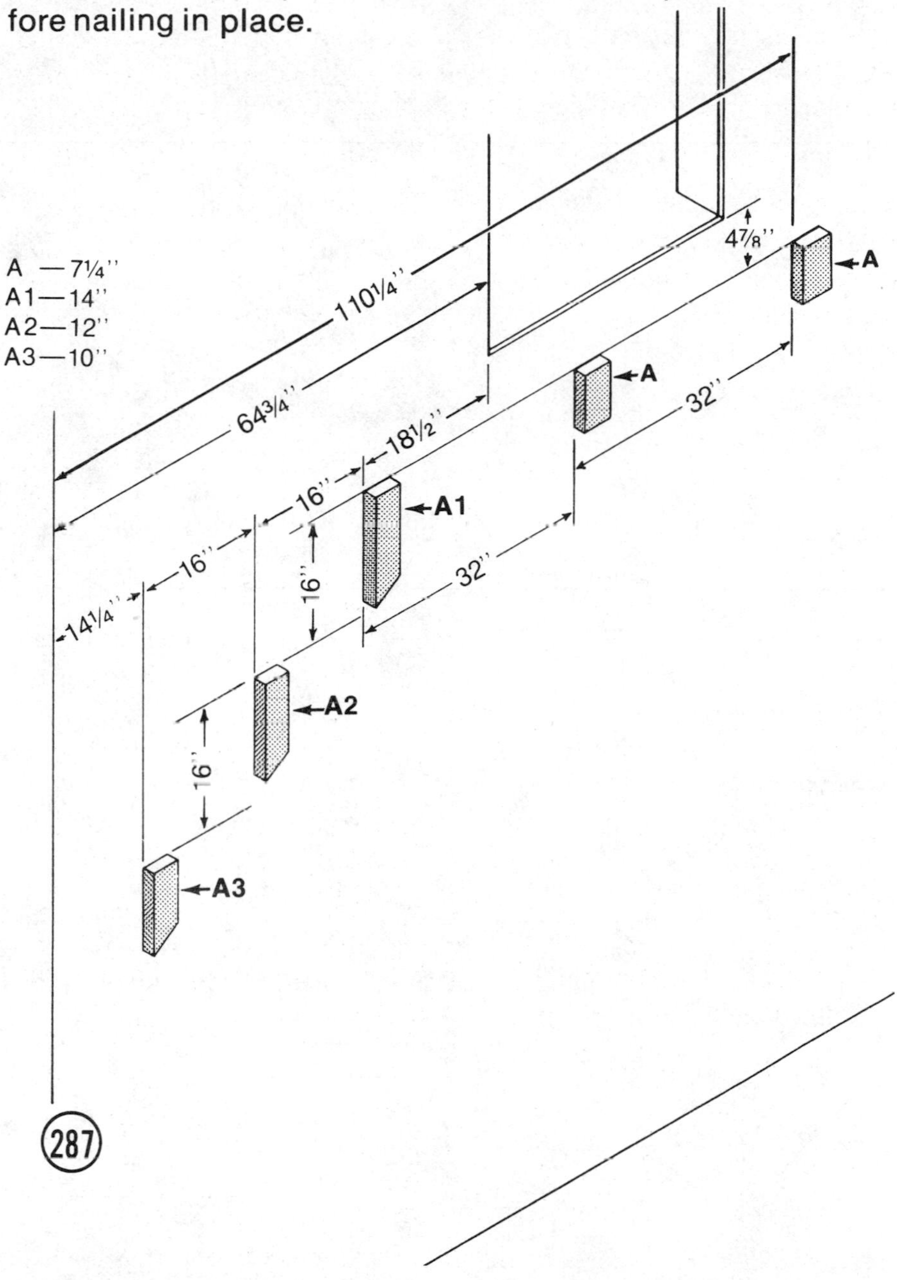

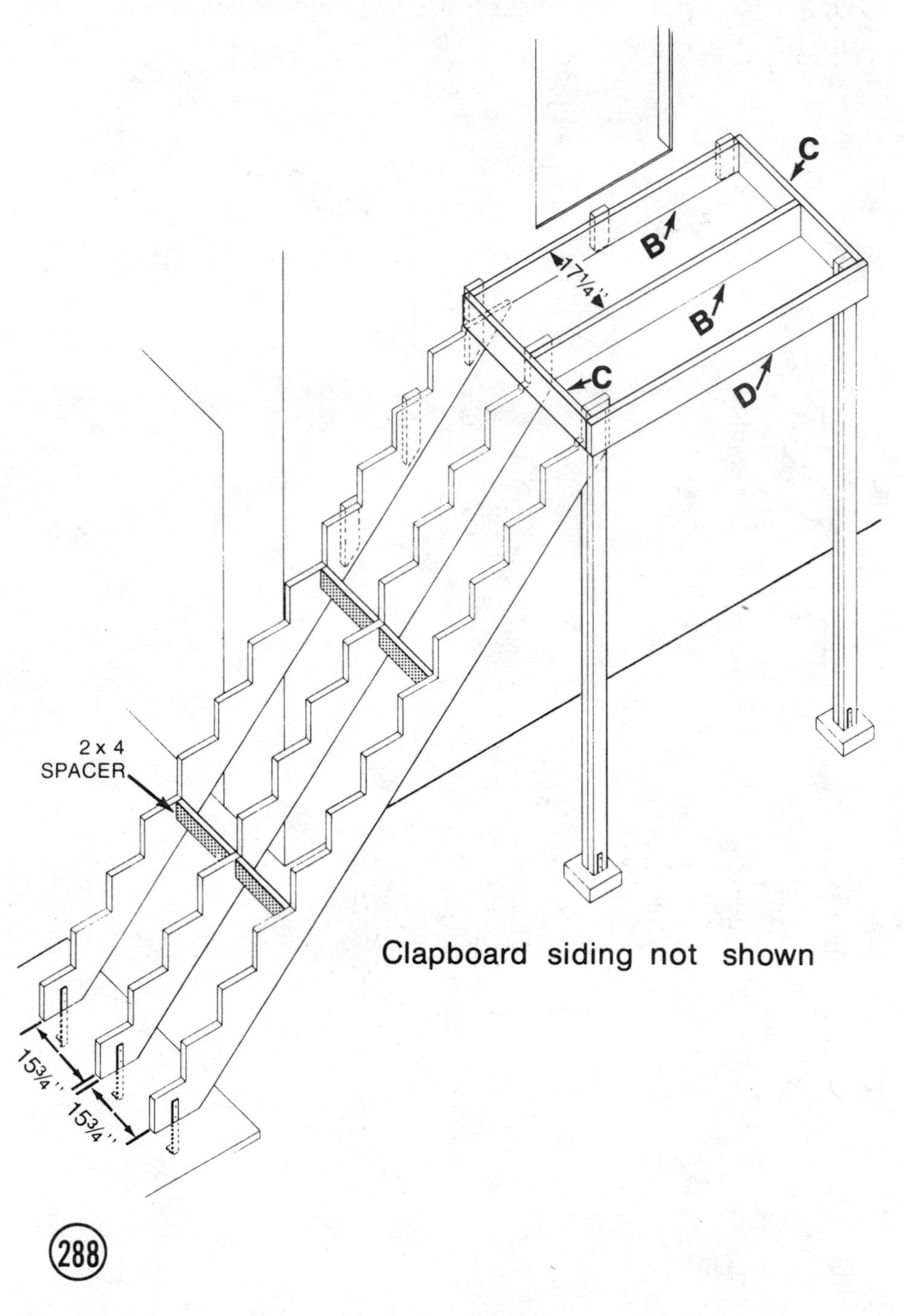
C
B
17¼"
B
C
D
2 x 4
SPACER
Clapboard siding not shown
15¾"
15¾"
288

Nail 2 x 4 x 7¼" furring blocks A, 2 x 4 x 14" A1, 2 x 4 x 12" A2, and 2 x 4 x 10" A3 in position indicated for stairs and platform measuring 41 x 70½", Illus. 288.

In new construction it's necessary to apply clapboard siding to stair wall after nailing furring blocks and before nailing carriage. Apply siding level with top of A. Balance of siding is applied after door and window frames are installed.

When installing stairs alongside an existing structure, use pieces of clapboard or shingle to fur out siding, Illus. 289.

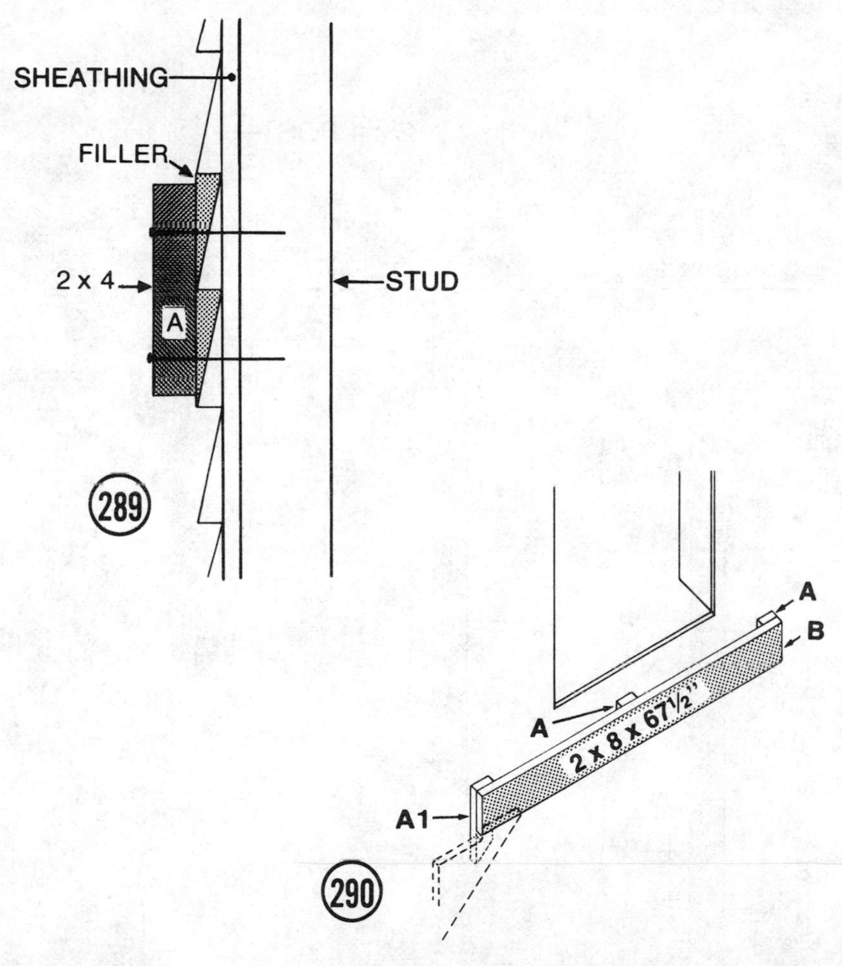

Nail one 2 x 8 x 67½" B in position 4⅞" below door opening, Illus. 290.

The average two story installation will require three 2 x 12 x 14' for carriages. You can get by with two by using a 2 x 4 for a center carriage, Illus. 258. Select straight lumber, free of knots.

Where 2 x 4 and 2 x 6 are used for treads, Illus. 291, the first riser should be cut 6½'' high. Follow procedure outlined on page 186. Place square 6½'' down instead of 6⅞'', Illus. 276, 277, 278. Draw position of each riser and tread.

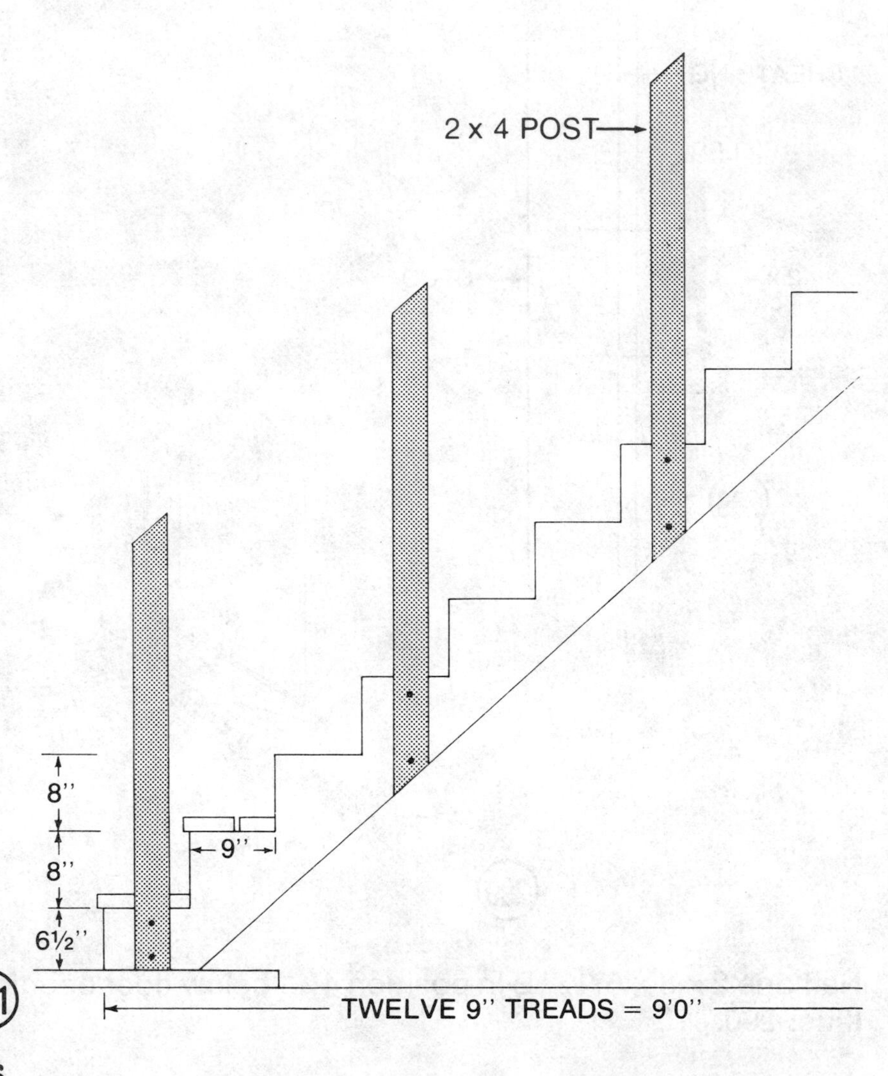

291

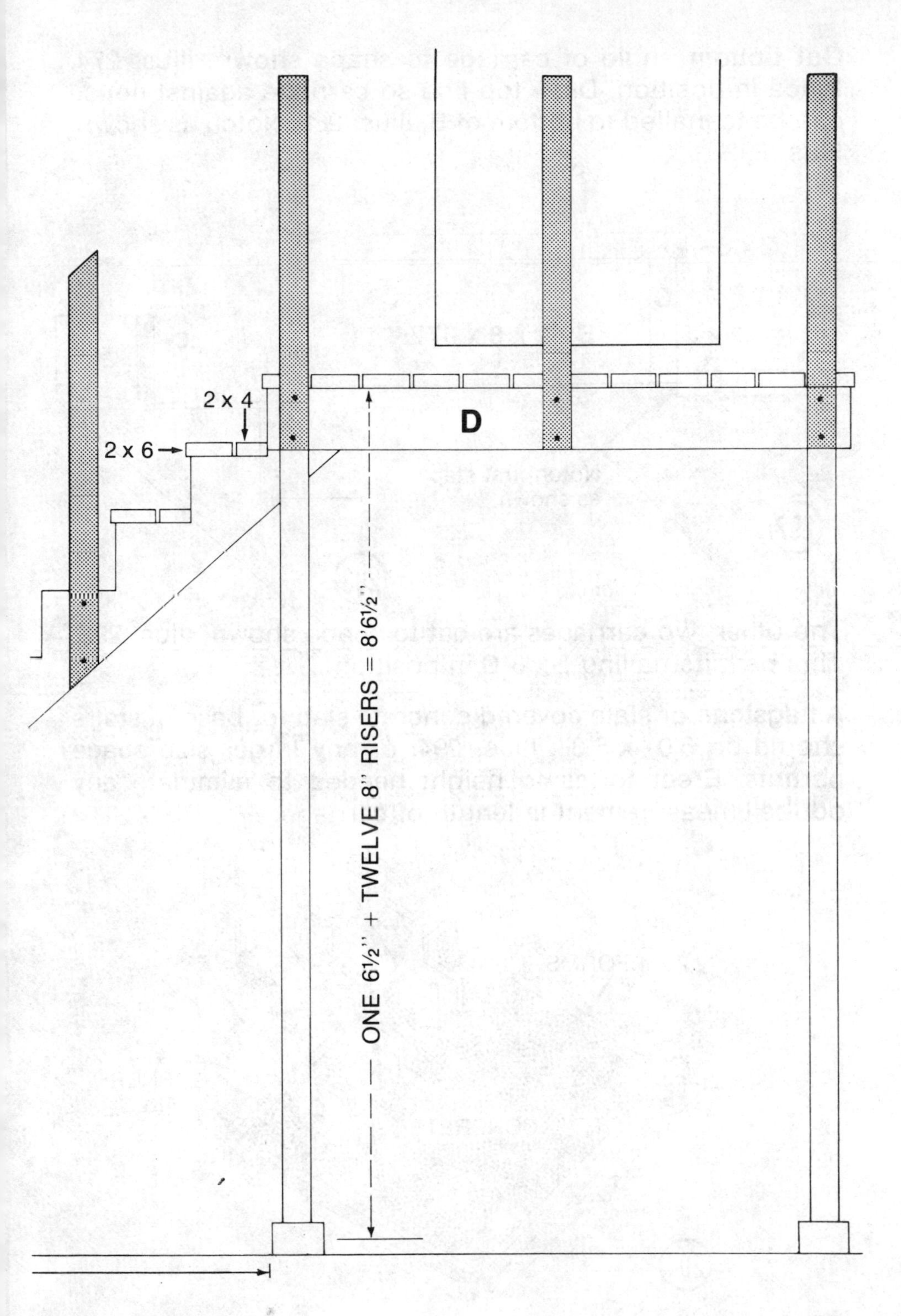
2 x 4
2 x 6
D
ONE 6½'' + TWELVE 8'' RISERS = 8'6½''

Cut bottom angle of carriage to shape shown, Illus. 274. Place in position. Draw top line so carriage against house can be toenailed to bottom of B, Illus. 288. Notch as shown, Illus. 292.

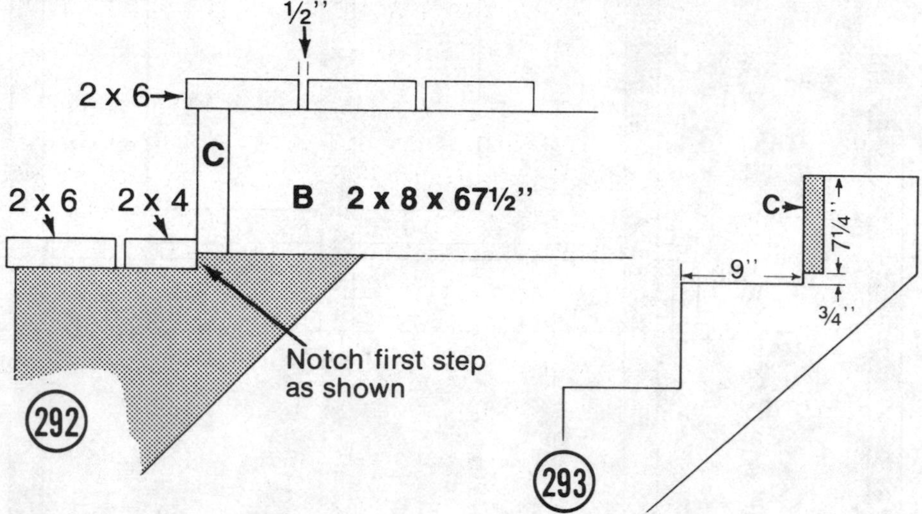

The other two carriages are cut to shape shown, Illus. 293. This permits nailing 2 x 8 C in position.

A flagstone or slate covered concrete slab for base of stairs should be 5'0" x 5'3", Illus. 294, or any larger size space permits. Erect forms to height needed to eliminate any oddball measurement in length of carriage.

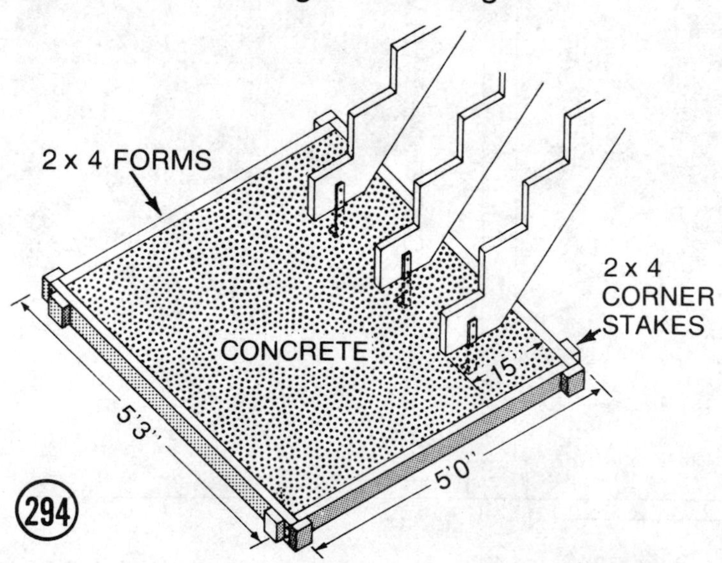

BUILD STAIR PLATFORM

After nailing 2 x 8 x 67½" B, Illus. 290, in position, cut and nail two 2 x 8 x 37½" C to B, Illus. 288. Cut one 2 x 8 x 70½" D, nail to C. Spike C to B in position shown. Check with level. Hold framing level with temporary 2 x 4's nailed to C.

Position 2 x 4 posts on concrete piers. Dig holes to depth below frost level. Allow form to project about an inch above grade. To locate position required, drop a plumb bob down from corner. Throw in some stone and set a 6" cardboard form, Illus. 131, in position at height above grade required. Or you can build a 6 x 6" form using 1 x 6. Fill and level form using a concrete mix. Embed a 12" length of 3/16 x 1" rustproof, painted steel strap in position post requires. Use a plumb bob to position strap.

Double check C and D to make certain both are level. Cut 2 x 4 outside post to length required. Plumb with level and spike to inside of D. Lag screw strap steel to base of post.

Spike first carriage to A1, A2 and A3. Toenail to bottom of B, Illus. 288. Spike center carriage to B. Spike third carriage to 2 x 4 post. Nail 2 x 4 spacers between carriages in position shown.

Cut inner 2 x 4 corner post, Illus. 295, to angle and length required. Toenail to carriage and to outer 2 x 4. Nail through C into carriage.

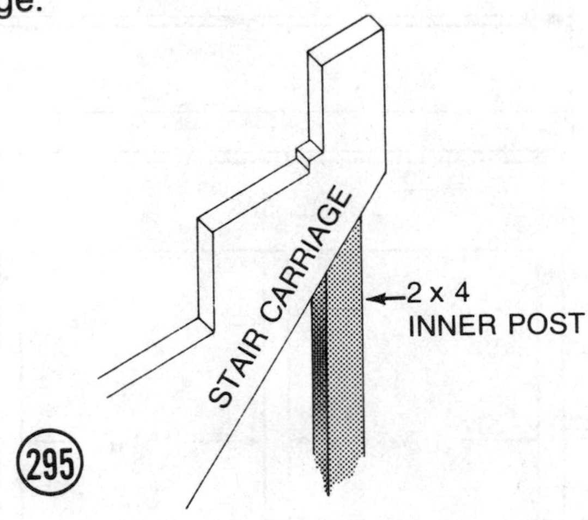

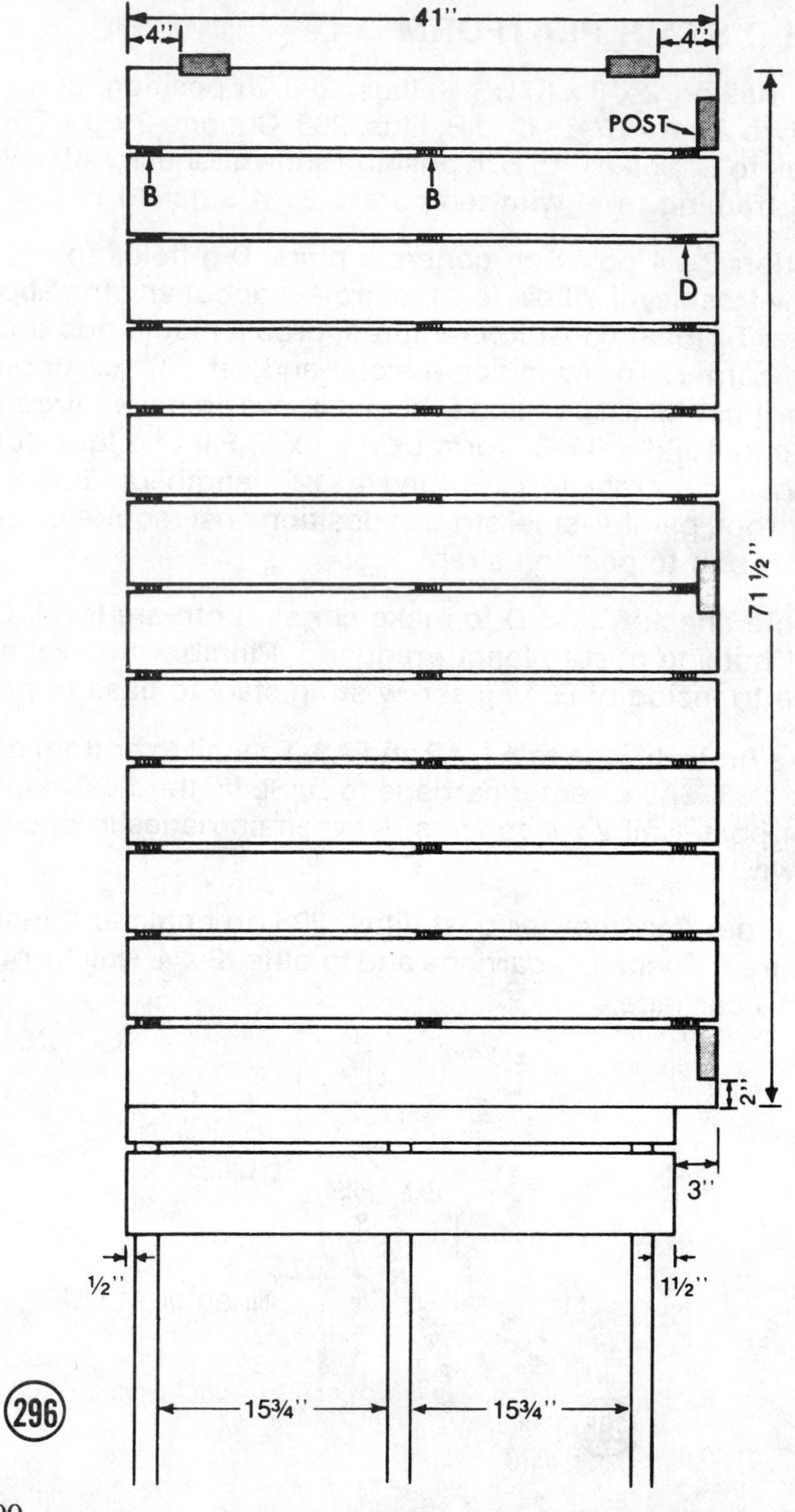
41''
4''
4''
POST
B
B
D
71 ½''
2''
3''
½''
1½''
15¾''
15¾''
296

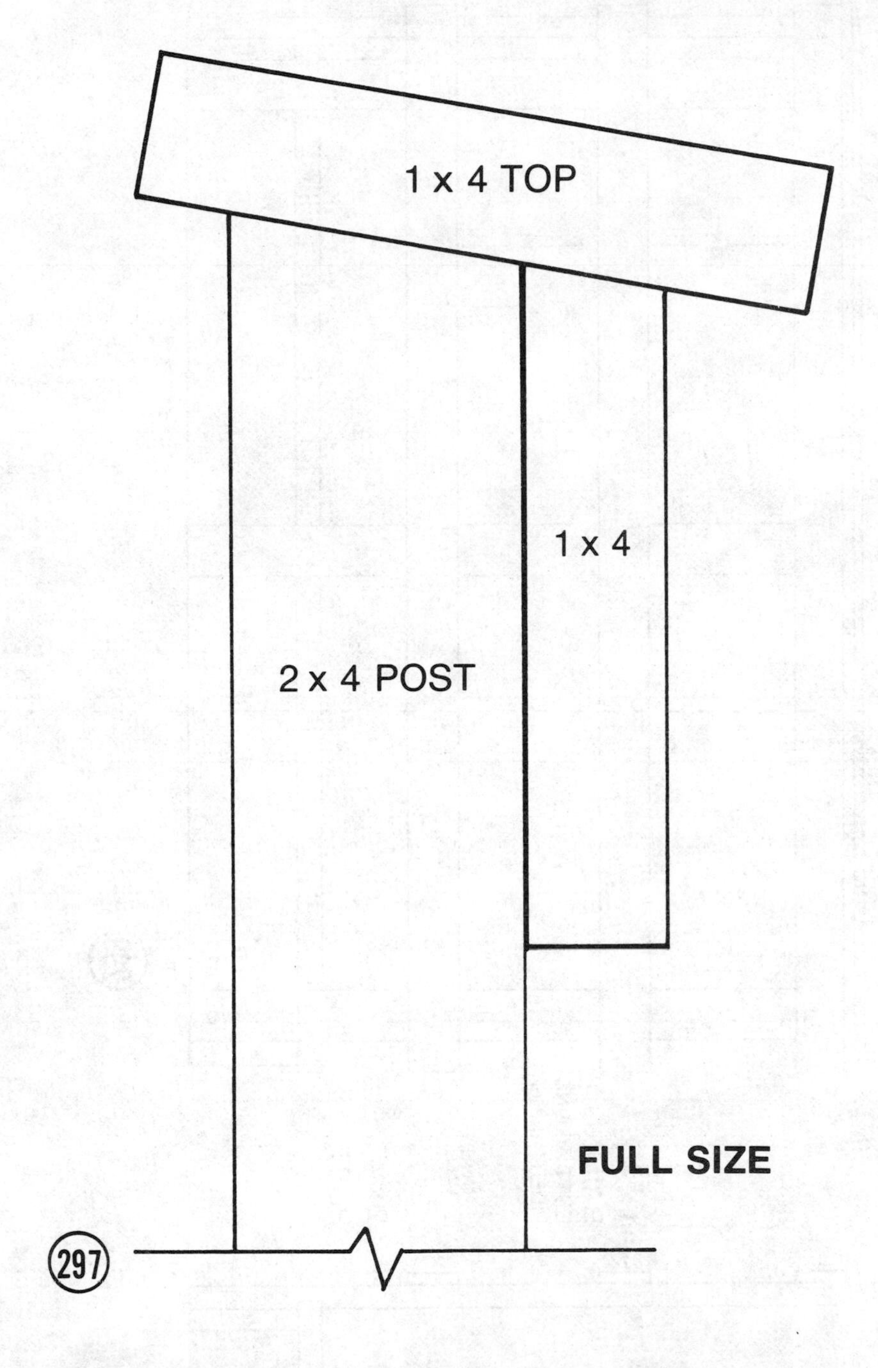

Illus. 296 shows location of 2 x 4 x 45'' platform railing posts.

Cut top to angle shown full size, Illus. 297.

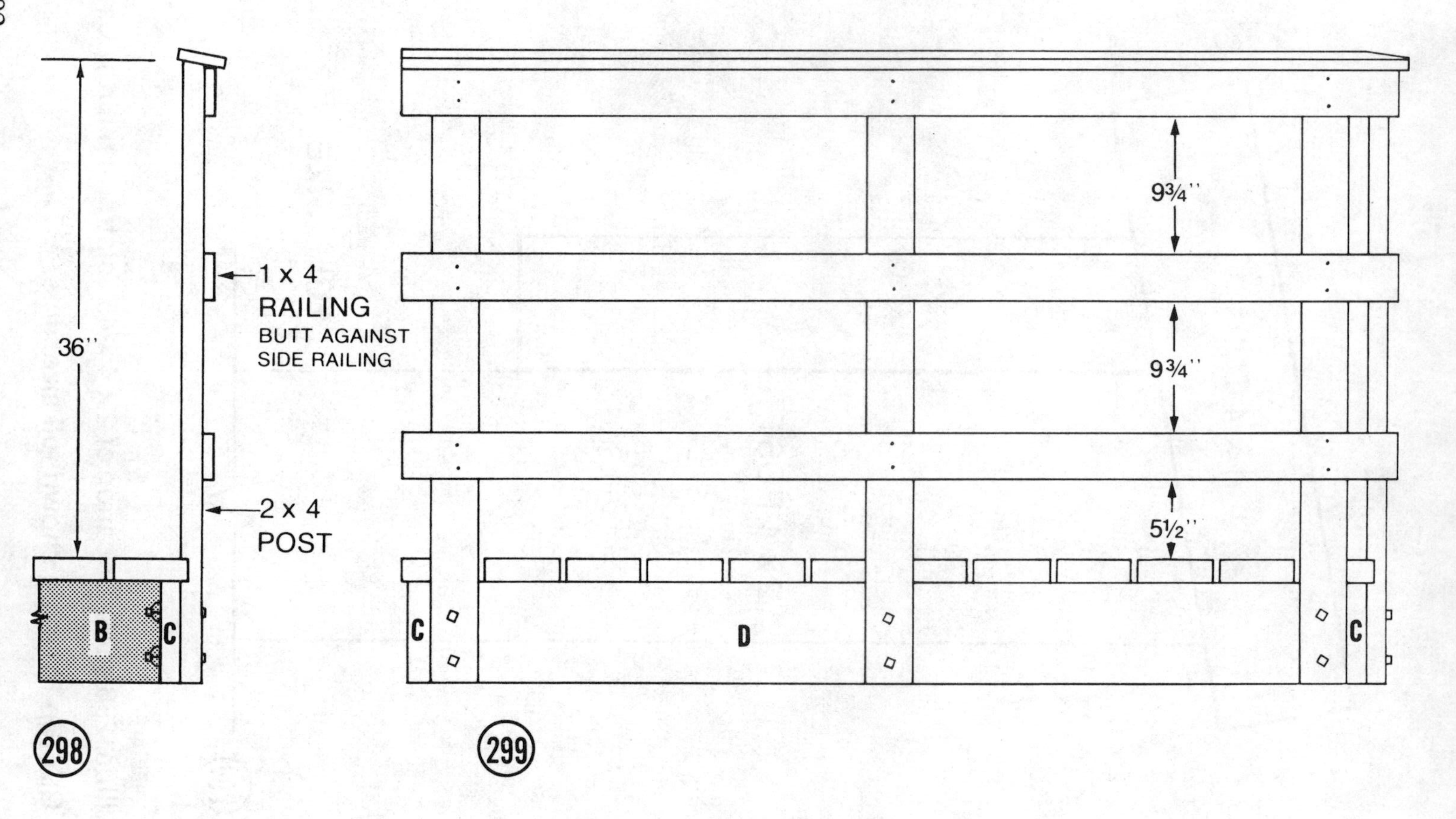

36''
1 x 4
RAILING
BUTT AGAINST
SIDE RAILING
2 x 4
POST
B
C
298
9¾''
9¾''
5½''
C
D
C
299

Bolt posts to C, D and to carriage with two ½ x 4" carriage bolts, Illus. 298, 296.

Spacing 2 x 6 platform flooring ½" apart requires twelve 2 x 6 x 41" for platform. Notch first 2 x 6 so it projects ½" over edge of C and 1½" over D, Illus. 296. Floor boards finish flush with post. Spike floor boards to BCD with 10 penny nails. You can cut two 2 x 6 x 41" and one 2 x 6 x 38" from a 2 x 6 x 10'.

Cut stair treads 2 x 6 x 38" and 2 x 4 x 38". Space ½" apart. Project tread ½" over carriage against house; 1½" on outside. Nail 2 x 4 and 2 x 6 in position indicated, Illus. 291.

Bolt 2 x 4 x 45" stair railing posts in position shown, Illus. 291. Always drill one hole. Bolt post in position. Plumb with level before drilling second hole through post and carriage.

Illus. 298 shows end view of 1 x 4 railing. Nail to posts with 8 penny finishing nails in position shown, Illus. 299. Use a file, round top edge slightly.

To miter cut top rail to angle required, cut a strip of wood to shape shown, Illus. 299A. Place in miter box in position shown. Saw 45° angle, Illus. 299B.

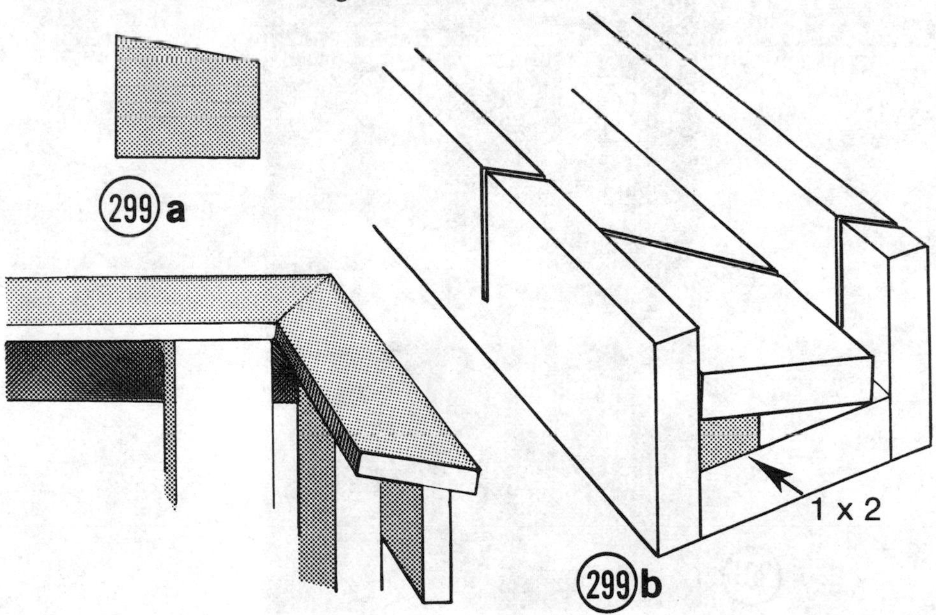

TO INSTALL AN OUTSIDE DOOR

Illus. 171 shows framing required for an outside door. Always frame opening to size retailer specifies. Nail furring blocks A in position to stiffen framing opposite hinges and strike plate. Use thickness plywood space allows, but don't force jamb.

To provide exact width jamb your construction requires, the retailer will have to know thickness of exterior sheathing, width of stud and thickness of interior paneling, Illus. 300. Filler strips can be used if necessary.

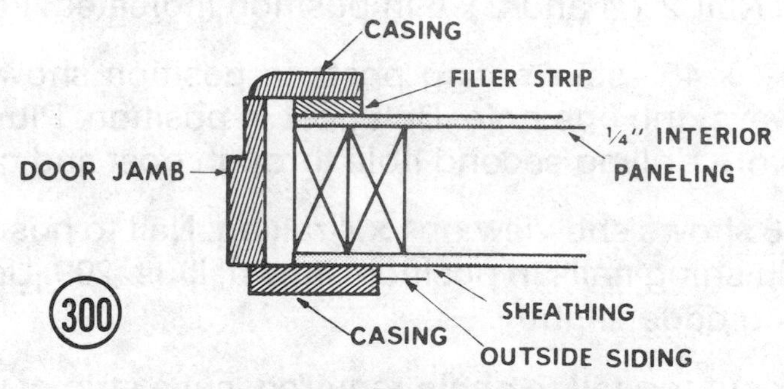

A prehung exterior door will usually have three pairs of 3½ x 3½" loose pin hinges. If you prefer a Dutch door, Illus. 301, these require four pairs of 4 x 4" brass, loose pin hinges. Buy an exterior door with small panes of double glass, or no glass. Install a one way viewer, Illus. 302.

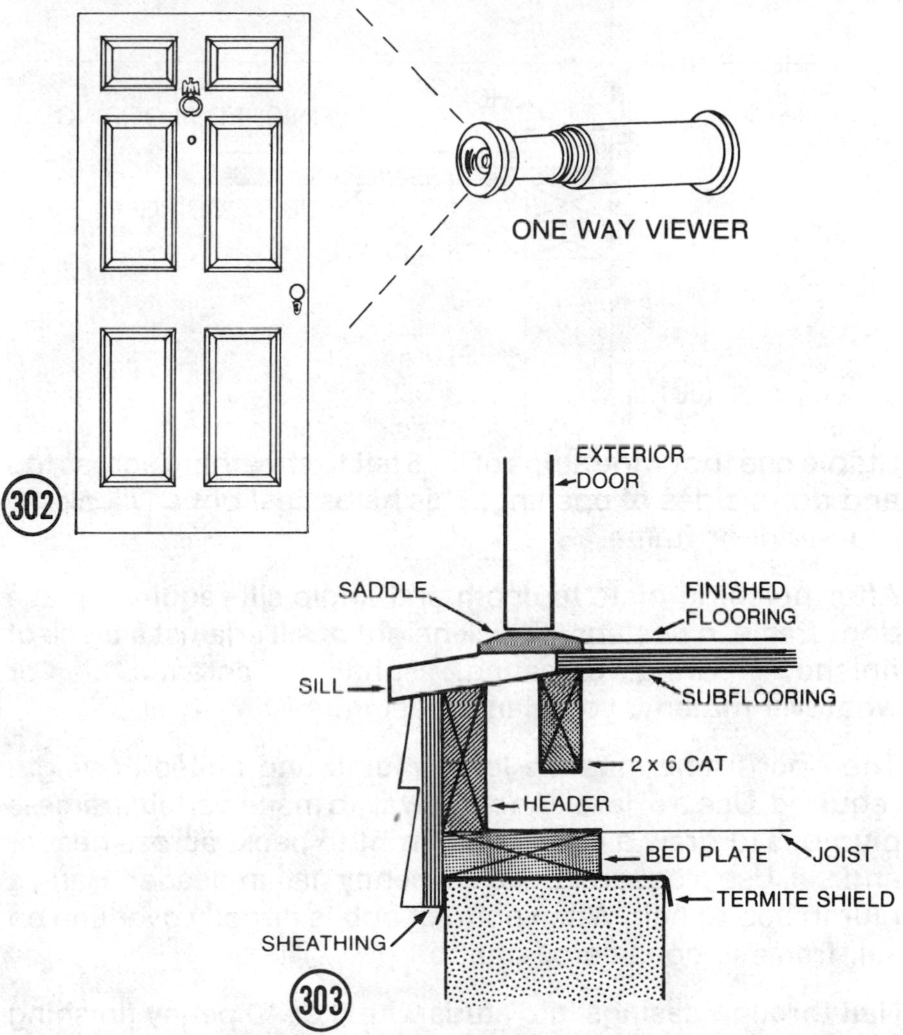

The first step is to cut shoe off across door opening, Illus. 171. The sill on an exterior door finishes slightly above finished flooring or carpeting. Since the door frame will be installed prior to finished flooring in new construction, always place a strip of finished flooring (or carpeting) in position, Illus. 303, to estimate position of sill.

Installing a door in a second story requires sawing through shoe, subflooring and plate, Illus. 304. Ask your dealer to advise exact depth and angle sill requires.

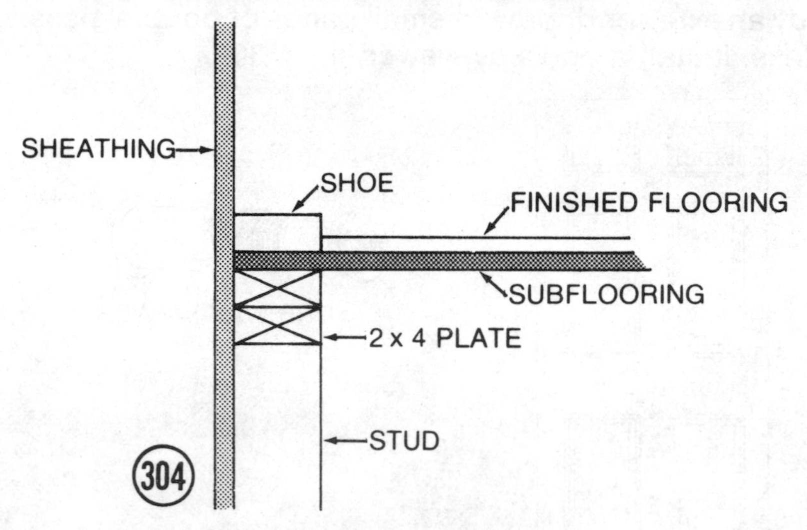

Staple one foot wide strips of #15 felt to sheathing across top and down sides of opening. This helps seal out any leakage around door frame.

After notching plate to depth and angle sill requires, place door frame in position. Check height of sill against a piece of finished flooring, carpeting, asphalt or ceramic tile, or whatever material you plan on using.

The door frame must be level, plumb and nailed at height required. Use a 6' level. Another way to make certain frame is plumb is to draw a center line, front to back, across header and sill. Use a square. Drive a 6 penny nail in header. Hang a plumb bob to nail. When point of bob is directly over line on sill, frame is considered plumb.

Nail through casings into studs with 8 or 10 penny finishing nails. Don't drive nails all the way. Test open door to make certain it swings freely and at a height finished floor covering requires. Prior to driving nails home, check corners with a square. Pack loose rock wool insulation in space around opening, then drive nails home. Countersink heads and fill holes with putty.

When installing an outside door in an existing building, saw clapboard siding off to width around opening equal to casings. Saw through sheathing and nail studs and header in position shown, Illus. 171. Use a double 2 x 4, 2 x 6 or size header prehung door frame requires.

Most pros use three hinges to install an exterior door, Illus. 305. One is placed 5" down from top, another 7" up from bottom; the third hinge at center. Hinges on Dutch doors are usually positioned to line up with panels.

To mortise a hinge in edge of door, pull hinge pin. Place leaf in position, Illus. 305. Mark outline on door. Using a 1" chisel, make cuts to depth thickness of leaf requires, Illus. 306.

Chisel out mortise, Illus. 307. Fasten leaf in position with screws provided.

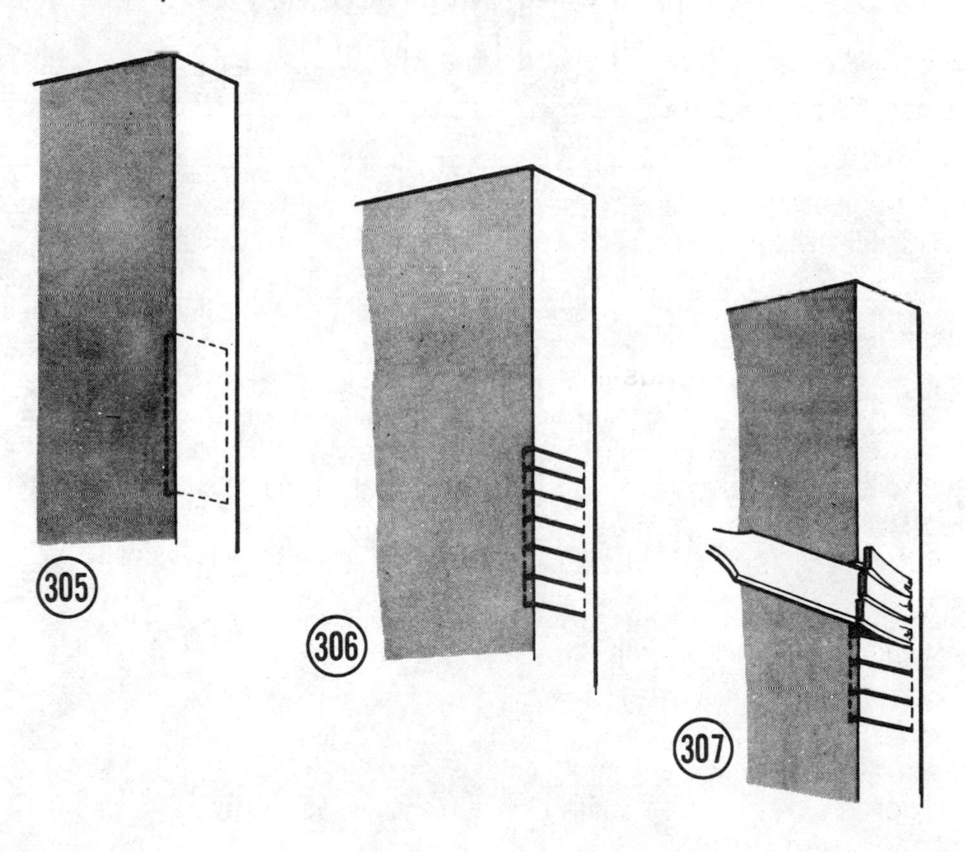

Follow same procedure to mortise out jamb, Illus. 308.

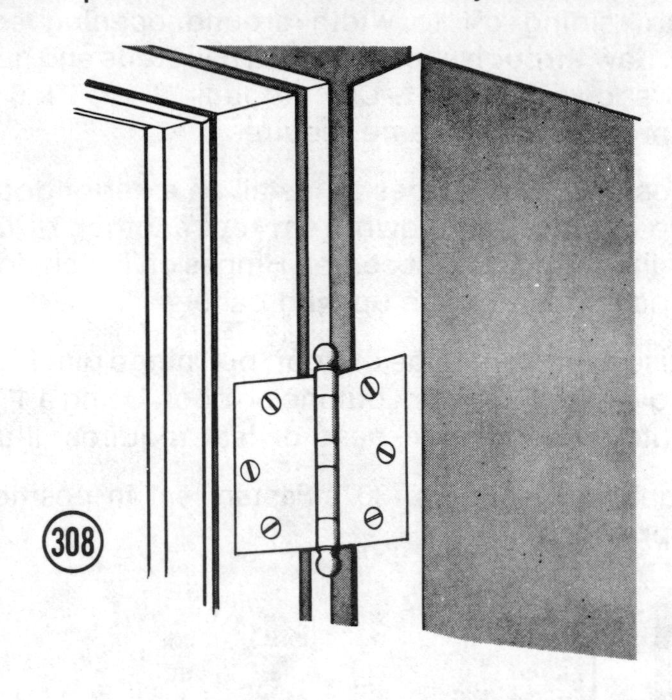

CONCRETE BLOCK SIZES

A– 7 5/8" B– 15 5/8" C–9 5/8" D– 11 5/8" E– 3 5/8"

APPROXIMATE METRIC SIZE A-19.37 B-39.7 C-24.45 D-29.53 E-9.21

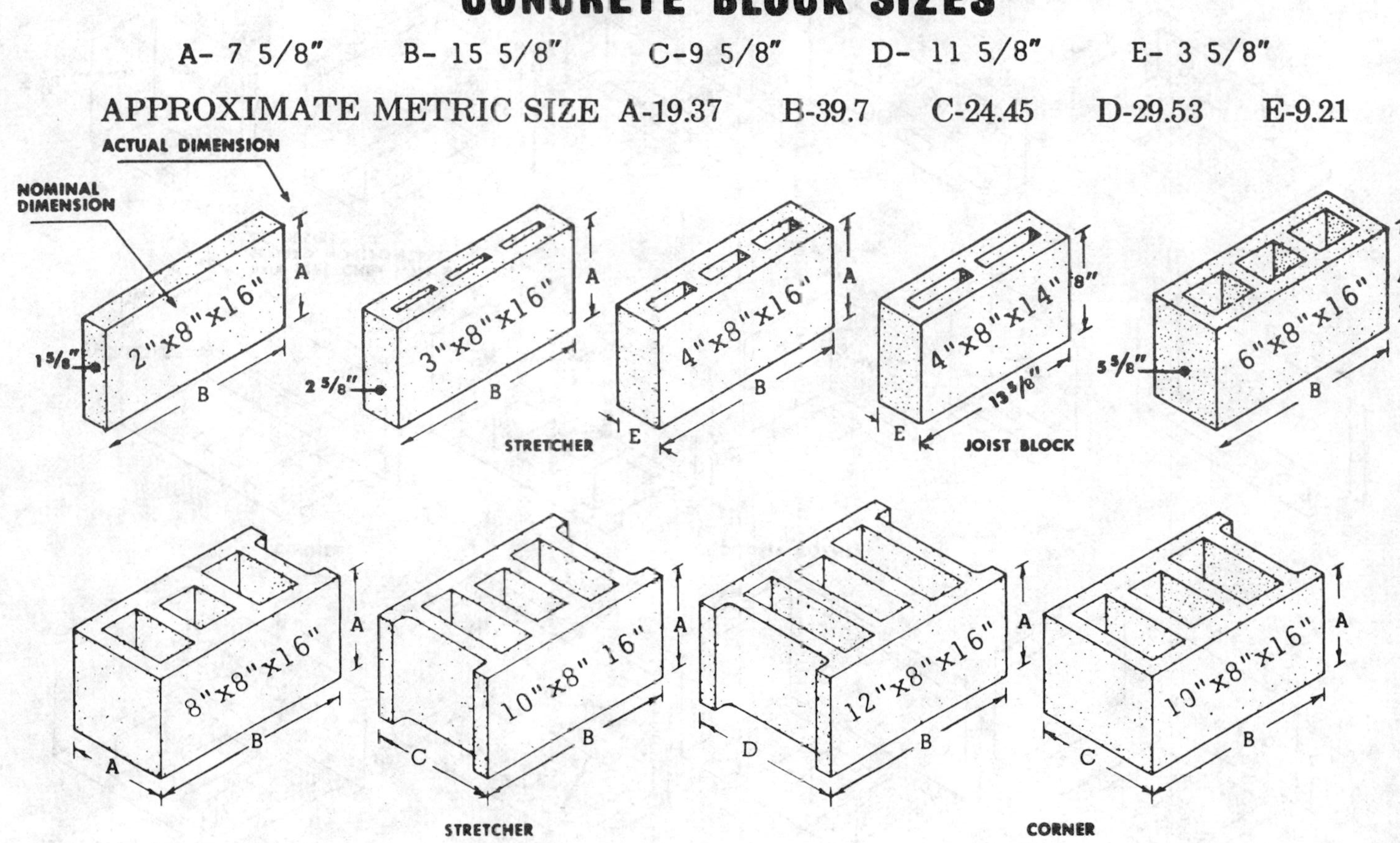

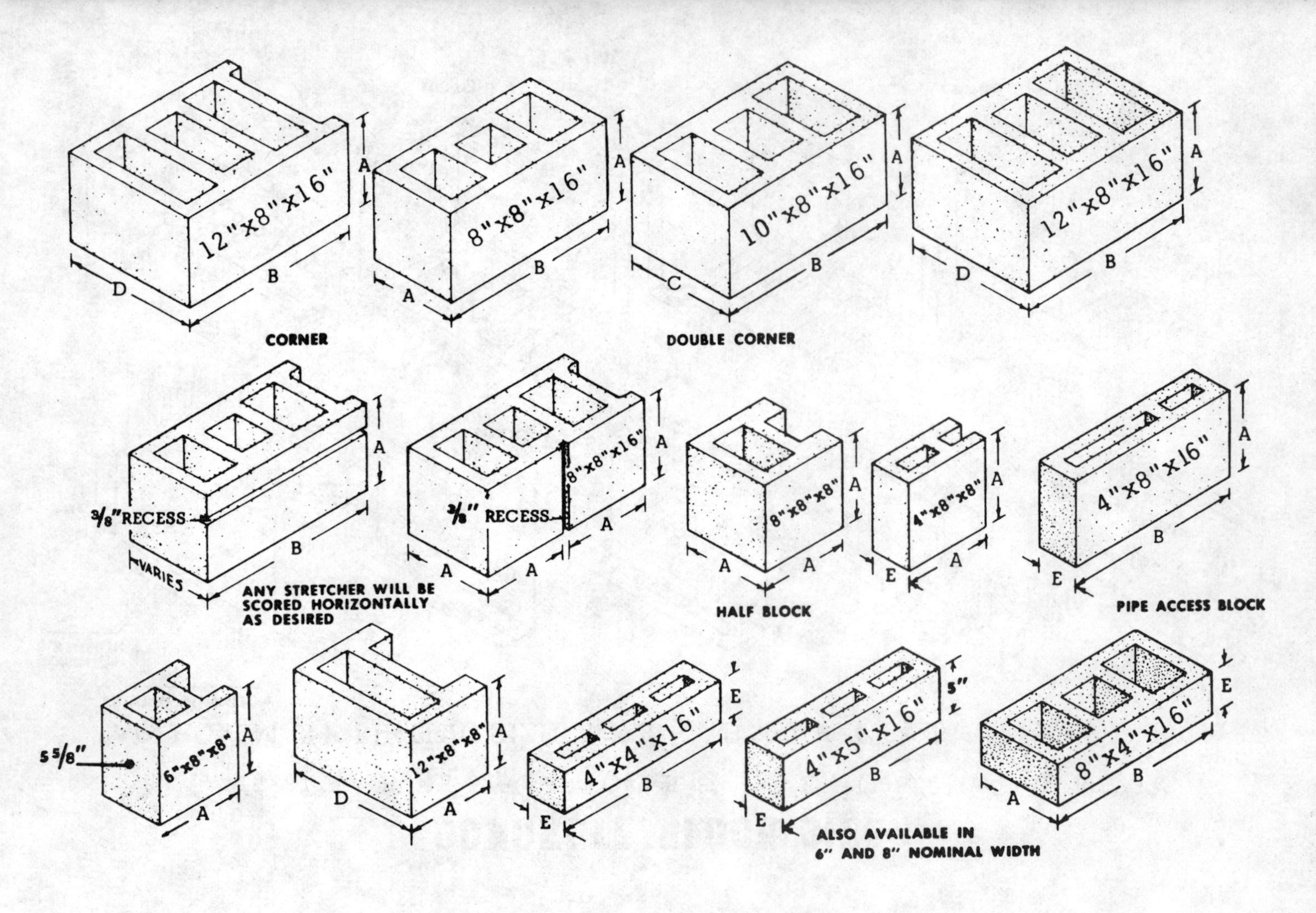

12"x8"x16"
8"x8"x16"
10"x8"x16"
12"x8"x16"
CORNER
DOUBLE CORNER
3/8"RECESS
VARIES
ANY STRETCHER WILL BE SCORED HORIZONTALLY AS DESIRED
3/8" RECESS
8"x8"x16"
8"x8"x8"
4"x8"x8"
HALF BLOCK
4"x8"x16"
PIPE ACCESS BLOCK
5 5/8"
6"x8"x8"
12"x8"x8"
4"x4"x16"
4"x5"x16"
5"
8"x4"x16"
ALSO AVAILABLE IN 6" AND 8" NOMINAL WIDTH

A STANDING OVATION FOR

BLESSINGS IN DISGUISE

"A most beguiling storyteller. . . . The men and women he writes about are the giants of the 20th century stage. He writes about them with such wit, candor and affection that he proves to be as splendidly gifted at creating characters on paper as he is on film."

—*Washington Post*

"A charming, readable and graceful work . . . Guinness writes with distinction, embellishing his memories with unusual insight and intimacy."

—*Chicago Tribune*

"The wittiest, most elegant memoir to come out of theatre in years."

—*London Sunday Times*

"The book is as delightful to read as he always was—is—to watch!"

—*New York Daily News*

"The last of the great English actors of our time to deliver his memoirs, Guinness has topped them all . . . *Blessings in Disguise* presents Guinness the actor, brilliantly at work."

—*Vanity Fair*

"An infinitely agreeable entertainment. . . . There won't be another actor's autobiography published for a long time that provides as much enjoyment as this book."

—*People*

"A wonderful book, engaging, ruthlessly funny."

—*Philadelphia Inquirer*

more . . .

"He is a warm and witty writer, and an enormously charming raconteur."
—*San Francisco Chronicle*

"Filled with childhood reminiscences, spiritual odysseys, wartime adventures, artistic puzzles and above all, with deft, piercingly honest sketches of great friendships. . . *Blessings in Disguise* proves that behind the greasepaint and the scripted lines lies a very real person, and to spend time in his company is an unqualified pleasure."
—*Chicago Sun-Times*

"A connoisseur's delight. . . .Likely to sail down to posterity with a twinkle in its sails, its ballast soundly laid on good writing and wit so dry that it sometimes threatens to evaporate."
—*Punch*

"Alec Guinness has few equals among English-speaking actors, and now he is discovered to be an uncommonly felicitous prose stylist as well."
—*Los Angeles Times*

A marvelous Guinness record. . . of a great period in the British theatre."
—Anthony Burgess, *Observer*

"A perfectly memorable memoir. . . irresistible. . . conjures up captivating portraits of John Gielgud, Tyrone Guthrie, Edith Evans and Ralph Richardson, with cameo appearances by Ernest Hemingway, Grace Kelly, Noel Coward, Vivian Leigh and Fidel Castro."
—*Houston Chronicle*

"Elegantly entertaining. . . suprisingly rich and rewarding."
—*Cleveland Plain Dealer*

"A blessing undisguised—the lively, literate reminiscences of one of the treasures of the British stage and screen. . . Guinness offers insights into the actor's art, thoughtful discussions of religious experience, encomiums to family life, plus deft portraits of everyone from Edith Sitwell to Sophia Loren, Pope Pius XII to Hemingway. . . a portrait of a wryly sensitive 'original.'"
—*Kirkus Reviews*